Common Core Advanced
Multiple Choice Question

RONALD V. COX
Gresham's School
Holt, Norfolk

Test Development and Research Unit
Objective Test Series

General Editor
John S. Hamilton

CAMBRIDGE
UNIVERSITY PRESS

Other books in this series
Multiple Choice Questions in Advanced Level Chemistry
Multiple Choice Questions in Advanced Level Biology

Published by the Press Syndicate of the University of Cambridge
The Pitt Building, Trumpington Street, Cambridge CB2 1RP
40 West 20th Street, New York, NY 10011-4211, USA
10 Stamford Road, Oakleigh, Victoria 3166, Australia

on behalf of the Oxford Delegacy of Local Examinations, the University of Cambridge Local Examinations Syndicate and the Test Development and Research Unit of the three G.C.E. boards associated with the Universities of Oxford and Cambridge.

© Cambridge University Press 1983

First published 1983
Fifth printing 1992

Printed in Great Britain at the
Athenaeum Press Ltd, Newcastle upon Tyne

Library of Congress catalogue card number: 82-14607

ISBN 0 521 27281 5

Contents

General Introduction		*page*	1
Introduction			2
Index to test items			3

Exercises

Section A	General physics and mechanics	5
Section B	Oscillations and waves	18
Section C	Electricity (i): d.c. circuits and electrostatics	30
Section D	Electricity (ii): electromagnetism, alternating current, charged particles	42
Section E	Molecular, thermal and mechanical properties of matter	57
Section F	Atomic and nuclear properties	69
Section G	Graphs	79
Section H	Errors	84

Item statistics and correct answers   86

General Introduction

This series arises from the work of the Test Development and Research Unit in the development and production of objective tests, and it is based on the belief that good objective tests have an educational as well as a measurement function. The Unit has always considered that a good test item (question) should not only possess the qualities of conciseness, clarity and originality, but that it should also be of a kind which teachers would like to have for teaching purposes and which students would find of real interest.

The test items included in this series originated as part of a larger number of items which were rigorously scrutinised, edited and pretested to ensure their relevance to the study of the subjects at the stated levels, and to permit the measures of their facility and discrimination (discussed elsewhere in this book) to be estimated in advance. Thus only the best of the original items are included in actual G.C.E. examinations, and it is from this stock of examination items that the present series has been compiled.

The items have two main uses for students. First, an objective test can provide a learning experience. Although most items in examinations are intended to discriminate effectively between candidates of different abilities, it is the examiners' policy to include some items which, although likely to yield comparatively low facility and discrimination indices, seem particularly useful in presenting novel situations and unusual problems. Such items expand the students' knowledge, whilst more familiar items consolidate the knowledge they already possess.

Second, and with the help of the statistics presented, it is possible to compare the performance of pre-examination students with that of examination candidates. If the abilities of individual students are to be evaluated reliably, however, a test containing at least thirty items should be used.

In preparing this series I have had much useful advice and criticism from experienced examiners and teachers, and from colleagues among the staffs of the Test Development and Research Unit and the examination boards. The editors of the individual books have taken great care in the task of presenting the best possible materials to students and teachers. Despite the complexities of editing and pretesting, our experience is that the most essential ingredient of good objective tests is the provision of good items as raw material, and the writing of objective test items is a skilled and often underrated task. Grateful acknowledgement is therefore made to all those who have contributed items to the boards' examinations, and I hope that this series will allow their efforts to be more widely appreciated.

John S. Hamilton
Test Development and Research Unit

Introduction

The multiple choice items in this book are taken from recent A-level Physics examinations set by the University of Cambridge Local Examinations Syndicate and the Oxford Delegacy of Local Examinations. The items chosen are those that correspond to the 'Core Syllabus', recommended in December 1980 by the Inter-Board Working Party in Advanced Level Physics, on which all G.C.E. boards were represented. This syllabus broadly outlines the minimum attainments for students who wish to embark upon studies involving physics at universities and polytechnics in the United Kingdom.

The booklet is arranged in eight sections. Each of sections A to F contains four Exercises, with the more complex items contained in Exercises 3 and 4. Each Exercise may be used as a whole, providing practice in varied topics. Alternatively, items on similar topics, at increasing stages of sophistication, will be found under the same item number in successive exercises: thus the four items numbered 1 in Exercises 1, 2, 3 and 4 of section A are all questions about 'Dimensions and units', and similar parallel arrangements will be found throughout the book. Sections G and H are short sections on 'Graphs' and 'Errors' respectively.

The earlier Exercises in each section are suitable for students during the elementary stages of A-level courses. Later Exercises often involve combinations of topics within single questions and are appropriate for the students' final preparations for the A-level examinations. An index to topics and items is provided (pp. 3/4) to assist both teachers and students in using this book to the best advantage.

All items are designed to be solved successfully by well-prepared candidates in a time not exceeding two minutes. Many of the items can be answered in a shorter time. Typical examination times are $1\frac{1}{4}$ hours for a 40-question test, and $1\frac{3}{4}$ hours for a 50-question test. The items are reproduced largely in the form in which they appeared in A-level examinations, except where modifications in terminology have been necessary in order to conform with current practice, as expressed in the 1981 edition of the Association for Science Education's booklet *SI Units, Signs, Symbols and Abbreviations*.

Ronald V. Cox

Index to test items

Section A   General physics and mechanics (pages 5 to 17)

Dimensions and units	A1.1	A2.1	A3.1	A4.1
Vector quantities	A1.2	A2.2	A3.2	A4.2
Kinematics	A1.3	A2.3	A3.3	A4.3
Momentum	A1.4	A2.4	A3.4	A4.4
Collisions	A1.5	A2.5	A3.5	A4.5
Work, power and energy	A1.6	A2.6	A3.6	A4.6
Circular motion	A1.7	A2.7	A3.7	A4.7
Gravitation	A1.8	A2.8	A3.8	A4.8
Satellite motion	A1.9	A2.9	A3.9	A4.9
Hydrostatics	A1.10	A2.10	A3.10	A4.10

Section B   Oscillations and waves (pages 18 to 29)

Simple harmonic motion	B1.1	B2.1	B3.1	B4.1
Mathematics of simple harmonic motion	B1.2	B2.2	B3.2	B4.2
Sound	B1.3	B2.3	B3.3	B4.3
Light waves and refraction	B1.4	B2.4	B3.4	B4.4
Interference of light waves	B1.5	B2.5	B3.5	B4.5
Diffraction	B1.6	B2.6	B3.6	B4.6
Diffraction gratings	B1.7	B2.7	B3.7	B4.7
Standing transverse waves	B1.8	B2.8	B3.8	B4.8
Standing longitudinal waves	B1.9	B2.9	B3.9	B4.9
Electromagnetic waves	B1.10	B2.10	B3.10	B4.10

Section C   Electricity (i): d.c. circuits and electrostatics (pages 30 to 41)

Resistance networks	C1.1	C2.1	C3.1	C4.1
Meter conversions	C1.2	C2.2	C3.2	C4.2
Electromotive force and internal resistance	C1.3	C2.3	C3.3	C4.3
Heat and power	C1.4	C2.4	C3.4	C4.4
Capacitors and capacitance	C1.5	C2.5	C3.5	C4.5
Capacitor networks	C1.6	C2.6	C3.6	C4.6
Capacitor energy	C1.7	C2.7	C3.7	C4.7
Capacitors in d.c. circuits	C1.8	C2.8	C3.8	C4.8
Electric potential	C1.9	C2.9	C3.9	C4.9
Electric field and forces	C1.10	C2.10	C3.10	C4.10

4 INDEX TO TEST ITEMS

Section D   Electricity (ii): electromagnetism, alternating current, charged particles (pages 42 to 56)

Magnetic flux	D1.1	D2.1	D3.1	D4.1
Electromagnetic forces	D1.2	D2.2	D3.2	D4.2
Electromagnetic induction	D1.3	D2.3	D3.3	D4.3
Inductance	D1.4	D2.4	D3.4	D4.4
Alternating current root-mean-square values	D1.5	D2.5	D3.5	D4.5
Diodes and rectification	D1.6	D2.6	D3.6	D4.6
Cathode-rays and oscilloscopes	D1.7	D2.7	D3.7	D4.7
Charged particles in magnetic fields	D1.8	D2.8	D3.8	D4.8
Charged particles in electric fields	D1.9	D2.9	D3.9	D4.9
Conduction processes	D1.10	D2.10	D3.10	D4.10

Section E   Molecular, thermal and mechanical properties of matter (pages 57 to 68)

Atomic and molecular magnitudes	E1.1	E2.1	E3.1	E4.1
Molecular properties	E1.2	E2.2	E3.2	E4.2
Molecular speeds	E1.3	E2.3	E3.3	E4.3
Gas laws	E1.4	E2.4	E3.4	E4.4
Temperature and thermometry	E1.5	E2.5	E3.5	E4.5
Heating, boiling and melting	E1.6	E2.6	E3.6	E4.6
Thermodynamics	E1.7	E2.7	E3.7	E4.7
Energy conversions	E1.8	E2.8	E3.8	E4.8
Thermal conduction	E1.9	E2.9	E3.9	E4.9
Elasticity	E1.10	E2.10	E3.10	E4.10

Section F   Atomic and nuclear properties (pages 69 to 78)

Photoelectric effect	F1.1	F2.1	F3.1	F4.1
Work function	F1.2	F2.2	F3.2	F4.2
Photons	F1.3	F2.3	F3.3	F4.3
Wave-particle duality	F1.4	F2.4	F3.4	F4.4
Energy transitions	F1.5	F2.5	F3.5	F4.5
Radioactivity	F1.6	F2.6	F3.6	F4.6
The nucleus	F1.7	F2.7	F3.7	F4.7
Nuclear reactions	F1.8	F2.8	F3.8	F4.8
Exponential decay	F1.9	F2.9	F3.9	F4.9
Half-life	F1.10	F2.10	F3.10	F4.10

Section G   Graphs G1-G10 (pages 79 to 83)

Section H   Errors H1-H10 (pages 84 to 85)

SECTION A   GENERAL PHYSICS AND MECHANICS

EXERCISE 1

A 1.1   Which one of the following units is dimensionally the same as the unit of acceleration?
  A ms^{-1} B Nm^{-1} C Ns^{-1} D Nkg^{-1}
  E Nms^{-1}

A 1.2   A body of mass 2.0 kg is moving on a horizontal surface with a speed of $1.41\,ms^{-1}$ in a north-easterly direction. A force of 0.20 N acting in a westerly direction is applied to the body for 10 s. If friction is negligible the body is then moving at
  A $0.41\,ms^{-1}$ in a north-easterly direction.
  B $1.00\,ms^{-1}$ in a northerly direction.
  C $1.41\,ms^{-1}$ in a north-westerly direction.
  D $2.24\,ms^{-1}$ in a direction 63.4° east of north.
  E $2.41\,ms^{-1}$ in a north-easterly direction.

A 1.3   Which one of the graphs below shows the distance s travelled after time t by a body starting from rest with uniform acceleration?

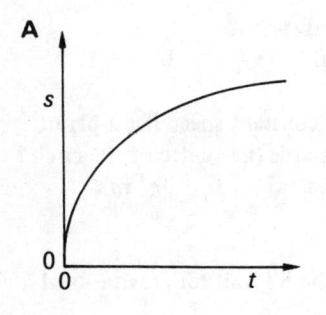

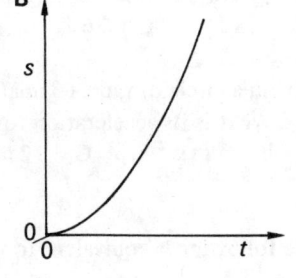

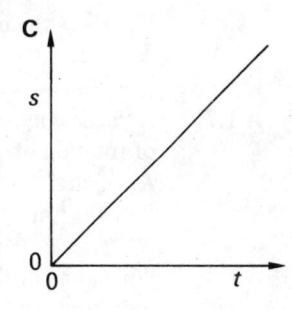

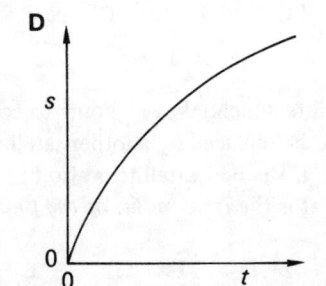

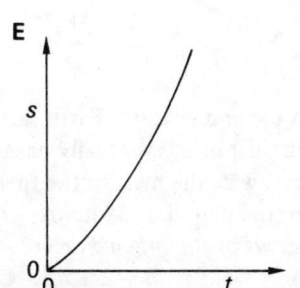

6   EXERCISE 1

A 1.4 A constant force acts on a body of mass m for time t, during which the speed of the body increases from u to v and the body travels a distance s. The change of momentum during this interval is given in magnitude by
A $m(v-u)/t$
B $m(v-u)/s$
C $m(v-u)$
D $m(v^2-u^2)/2$
E $m(v^2-u^2)/2s$

A 1.5 A body of mass m travelling with speed $5u$ collides with and adheres to a body of mass $5m$ travelling in the same direction with speed u. The speed with which the two travel together in the original direction is
A zero   B $\frac{8}{10}u$ C u D $\frac{6}{5}u$ E $\frac{10}{6}u$

A 1.6 In the diagram below the trucks X and Y are held in the position shown with the spring compressed. When they are released and the spring has fallen away, X is found to be moving to the left at $4\,\text{m s}^{-1}$.

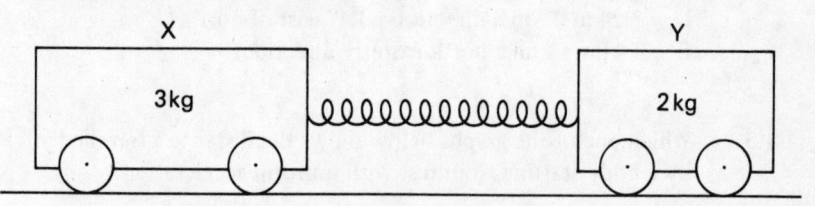

What was the energy stored in the compressed spring?
A 12 J   B 24 J   C 36 J   D 48 J   E 60 J

A 1.7 A particle moving in a circle of radius 3 m at constant speed has a period of rotation of 2 s. What is its acceleration towards the centre of the circle?
A $\frac{3}{4}\,\text{m s}^{-2}$ B $\frac{4}{3}\,\text{m s}^{-2}$ C $12\,\text{m s}^{-2}$ D $3\pi^2\,\text{m s}^{-2}$
E $48\pi^2\,\text{m s}^{-2}$

A 1.8 Which one of the following is equivalent to the SI unit for gravitational field strength?
A m s^{-2} B J kg^{-1} C kg N^{-1} D N m^{-1}
E $\text{N kg}^{-2}\,\text{m}^2$

A 1.9 A communications Earth satellite which takes 24 hours to complete one circular orbit eventually has to be replaced by another satellite which has twice the mass of the first. If the new satellite is also to have an orbital period of 24 hours, what is the ratio *radius of the final orbit*: *radius of the original orbit*?
A 1:1   B 2:1   C $2^{\frac{1}{2}}:1$ D 1:2   E $1:2^{\frac{2}{3}}$

A 1.10 An incompressible liquid of density ρ is contained in a vessel of uniform cross-sectional area A. If the atmospheric pressure is p and the acceleration due to gravity g has a constant value, then the downward force acting on a horizontal plane of area a situated at a depth d in the liquid is given by

A $Ap + ad\rho g$ B $\dfrac{p}{A} + \dfrac{d\rho g}{a}$ C $\dfrac{p + d\rho g}{a}$

D $\dfrac{a}{p + d\rho g}$ E $a(d\rho g + p)$

EXERCISE 2

A 2.1 Which one of the following quantities has the same dimensions as impulse?
- A   kinetic energy
- B   momentum
- C   potential energy
- D   weight
- E   work

A 2.2 A passenger in an open car travelling at 30 m s^{-1} throws a ball out over the bonnet. Relative to the car, the initial velocity of the ball is 20 m s^{-1} at 60° to the horizontal. What is the angle of projection of the ball with respect to the horizontal road?
A   23.4°   B   33.7°   C   40.0°   D   56.3°
E   66.5°

A 2.3 A body moves from rest with a constant acceleration. Which one of the following graphs represents the variation of its kinetic energy E_k with the distance travelled s?

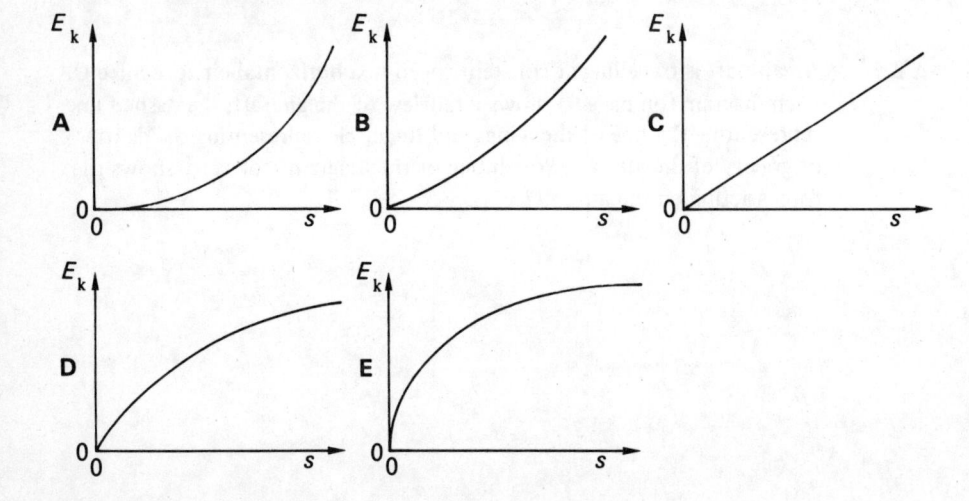

EXERCISE 2

A 2.4 A body of mass 3 kg is acted on by a force which varies as shown in the graph below.

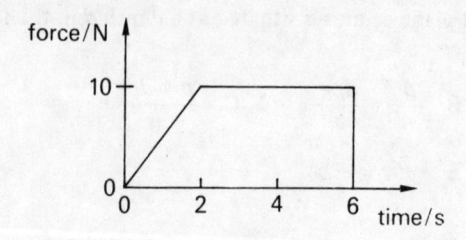

force/N vs time/s: rises from 0 at t=0 to 10 at t=2, constant at 10 until t=6, then drops to 0.

The momentum acquired is

A   0   B   5 N s   C   30 N s   D   50 N s   E   60 N s

A 2.5 Particles X (of mass 4 units) and Y (of mass 9 units) move directly towards each other, collide, and then separate.
If Δv_X is the change in the velocity of X, and Δv_Y the change in the velocity of Y, what is the magnitude of the ratio $\Delta v_X/\Delta v_Y$?

A   9/4   B   3/2   C   1/1   D   2/3   E   4/9

A 2.6 A bullet of mass 1.0×10^{-2} kg travelling horizontally at 500 m s^{-1} strikes and becomes embedded in, a stationary wooden block of mass 1.0 kg suspended by cords so that it can swing freely. The vertical height to which block and bullet swing is approximately

A   0.25 m   B   0.5 m   C   1 m   D   1.25 m
E   2.5 m

A 2.7 An aircraft is travelling at constant speed in a horizontal circle, centre Q. Each diagram (on page 9) shows a tailview of the aircraft, the dashed line representing the line of the wings and the circle representing the centre of gravity of the aircraft. Which one of the diagrams correctly shows the forces acting on the aircraft?

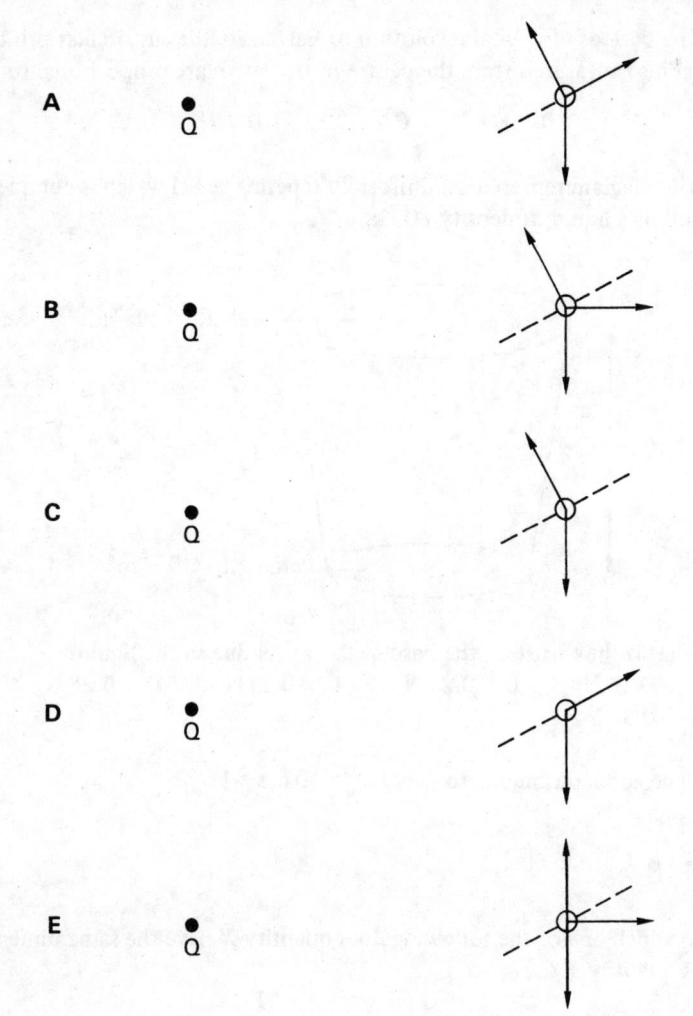

A 2.8 The values of the acceleration of free fall, g, on the surfaces of two planets are the same provided that the planets have the same
   A   mass
   B   radius
   C   mass/radius
   D   mass/(radius)^2
   E   mass/(radius)^3

EXERCISE 2

A 2.9 The periods of orbital revolution of Earth-satellites in circular orbits of radius r measured from the centre of the Earth are proportional to

A r^{-3} B $r^{-\frac{3}{2}}$ C $r^{-\frac{1}{2}}$ D $r^{\frac{2}{3}}$ E r^3

A 2.10 The diagram represents a uniformly tapering vessel which is completely full of a liquid of density 800 kg m^{-3}.

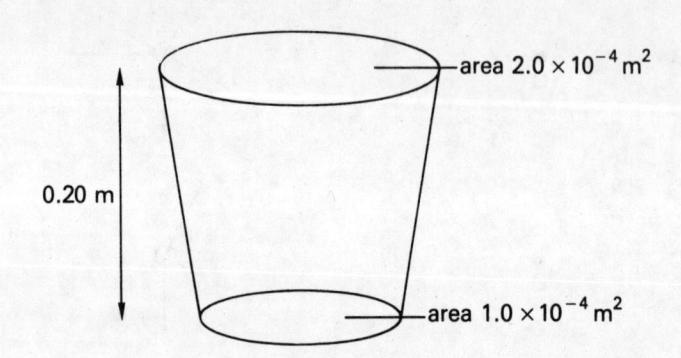

[Diagram: tapering vessel, height 0.20 m, top area $2.0 \times 10^{-4} \text{ m}^2$, bottom area $1.0 \times 10^{-4} \text{ m}^2$]

What is the thrust on the base of the vessel due to the liquid?

A 0.16 N   B 0.20 N   C 0.24 N   D 0.28 N
E 0.32 N

[The acceleration due to gravity $g = 10 \text{ m s}^{-2}$.]

EXERCISE 3

A 3.1 In which one of the following does quantity X have the same dimensions as quantity Y?

	X	Y
A	momentum/wavelength	energy/frequency
B	momentum × wavelength	energy/frequency
C	momentum/wavelength	energy × frequency
D	momentum/wavelength	frequency/energy
E	momentum × wavelength	energy × frequency

A 3.2 A horizontal force F is applied to a body of mass m on a smooth plane inclined at an angle θ to the horizontal, as shown below.

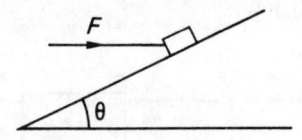

What is the magnitude of the resultant force acting on the body?

A $F\cos\theta - mg\sin\theta$
B $F\sin\theta + mg\cos\theta$
C $F\sin\theta - mg\cos\theta$
D $F\cos\theta + mg\sin\theta$
E $F + mg\tan\theta$

[g is the acceleration due to gravity.]

A 3.3 A tennis ball is released so that it falls vertically to the floor and bounces back again. Taking velocity upwards as positive, which of the following graphs best represents the variation of its velocity v with time t?

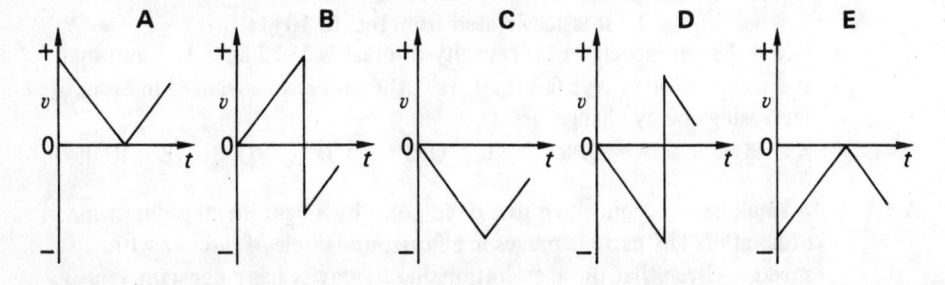

A 3.4 A projectile of mass m is fired with velocity v from a point P, as shown below.

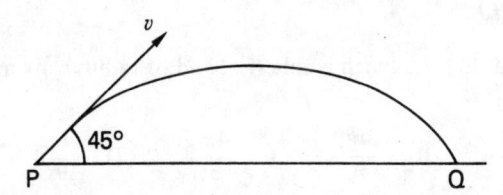

Neglecting air resistance, what is the magnitude of the change in momentum between leaving P and arriving at Q?

A zero   B $\frac{1}{2}mv$ C $\sqrt{2}mv$ D mv E $2mv$

EXERCISE 3

A 3.5 A vehicle P is moving without friction on a horizontal linear air-track with speed v and is about to collide with a stationary vehicle Q of the same mass, as illustrated below.

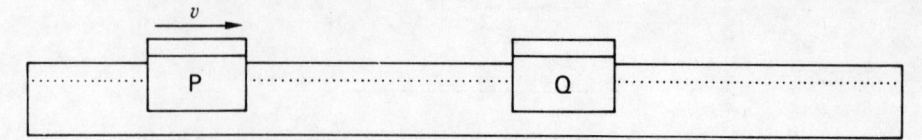

If the collision is elastic, then after the collision

A   P and Q move on together with speed $\tfrac{1}{2}v$.
B   P and Q move on together with speed $(1/\sqrt{2})v$.
C   P stops while Q moves on with speed v.
D   P rebounds with speed v while Q remains at rest.
E   P rebounds with speed $(1/\sqrt{2})v$ while Q moves on with speed $(1/\sqrt{2})v$.

A 3.6 A piece of brass undergoes three different processes involving change of energy:
    P: it is lifted vertically 2 m
    Q: it is heated from 15 °C to 20 °C
    R: it is accelerated from rest to 10 m s^{-1}.
Given that the specific heat capacity of brass is $380 \text{ J kg}^{-1} \text{ K}^{-1}$ and that the acceleration of free fall is 10 m s^{-2}, the processes, arranged in order of increasing energy change, are

A   PQR   B   QPR   C   QRP   D   PRQ   E   RQP

A 3.7 A small body is connected to a fixed point by a light inextensible string of length l. The particle moves in a horizontal circle of radius r with speed v. Given that the acceleration due to gravity has a constant value g, the angle α that the string makes with the vertical is such that $\tan \alpha$ is

A (v^2/gl) B (v^2/gr) C (gr/v^2) D (gl/v^2)
E (rv^2/g)

A 3.8 Assuming that the Earth is spherical and of radius r, its mean density is given by

A $\dfrac{4\pi rG}{3g}$ B $\dfrac{3rg}{4\pi G}$ C $\dfrac{4\pi rg}{3G}$ D $\dfrac{4\pi g}{3rG}$ E $\dfrac{3g}{4\pi rG}$

[g is the acceleration due to gravity at the Earth's surface; G is the gravitational constant.]

A 3.9 Two satellites S_1 and S_2 describe circular orbits of radii r and $2r$ respectively around a planet. If the orbital angular velocity of S_1 is ω, what is that of S_2?

A $\dfrac{\omega}{2\sqrt{2}}$ B $\dfrac{\omega\sqrt{2}}{3}$ C $\dfrac{\omega}{2}$ D $\dfrac{\omega}{\sqrt{2}}$ E $\omega\sqrt{2}$

A 3.10 A U-tube with its limbs vertical is half-filled with water of density 1000 kg m^{-3}. Paraffin oil of density 800 kg m^{-3} is then poured into one of the limbs to form a column of length 0.05 m. To what height will the water in the other limb rise above its original level?

A 0.005 m   B 0.01 m   C 0.02 m   D 0.025 m
E 0.033 m

EXERCISE 4

A 4.1 The stress σ required to fracture a solid can be expressed as

$$\sigma = k\sqrt{\left(\frac{\gamma E}{d}\right)},$$

where k is a dimensionless constant, E is the Young modulus and d is the distance between the planes of atoms separated by the fracture. Which one of the following quantities corresponds to γ, if the equation is dimensionally consistent?

A an energy per unit area
B a force per unit area
C a force
D an energy
E a density

A 4.2 A projectile is fired with an initial velocity u at an angle θ to the horizontal, as shown below, in a region where the acceleration due to gravity has a constant value g.

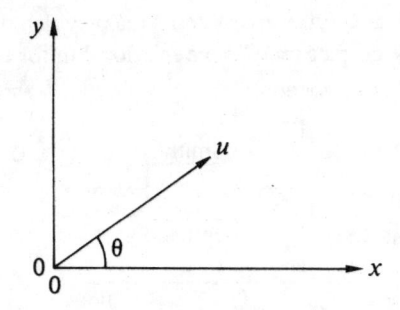

Neglecting air resistance, what is the height y and what is the horizontal distance x the projectile has travelled at time t after projection?

A $y = ut\cos\theta - \frac{1}{2}gt^2$ $x = ut\sin\theta$
B $y = ut\sin\theta - \frac{1}{2}gt^2$ $x = ut\cos\theta + \frac{1}{2}gt^2$
C $y = ut\cos\theta + \frac{1}{2}gt^2$ $x = ut\sin\theta$
D $y = ut\cos\theta$ $x = ut\sin\theta - \frac{1}{2}gt^2$
E $y = ut\sin\theta - \frac{1}{2}gt^2$ $x = ut\cos\theta$

EXERCISE 4

A 4.3 A parachutist steps from an aircraft, falls freely for 2 seconds, and then opens his parachute. Which of the following acceleration-time (a-t) graphs best represents his downwards acceleration a during the first 5 seconds of his motion?

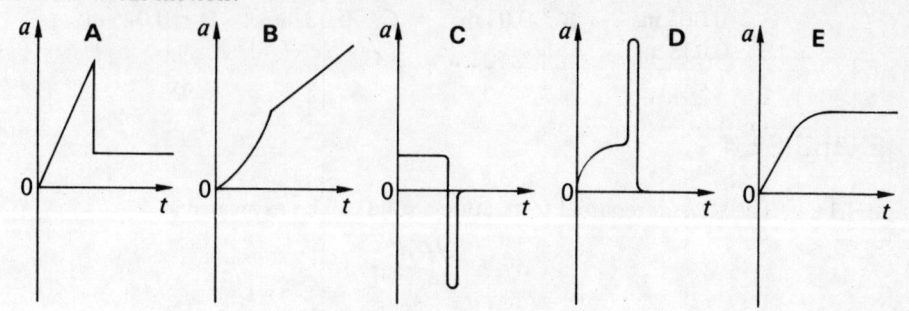

A 4.4 A stationary rocket of initial mass m discharges a jet of mean density ρ and effective area A at velocity v. The minimum value of v, which enables the rocket to rise vertically at a place where the acceleration due to gravity has a constant value g, is approximately

A $\sqrt{\left(\dfrac{2\pi g^2}{A}\right)}$ B $\sqrt{\left(\dfrac{\pi g}{Am}\right)}$ C $\sqrt{\left(\dfrac{\rho g A^2}{m}\right)}$

D $\sqrt{\left(\dfrac{mg}{A\rho}\right)}$ E $\sqrt{\left(\dfrac{2Amg}{\rho}\right)}$

A 4.5 Smith and Jones are skating on ice (assumed frictionless) so that they are moving with equal speed v in the same straight line. Smith is skating backwards facing Jones. Smith throws a ball to Jones at time t_1 and receives it back at time t_2. Assuming that the time of flight of the ball is negligible and that Smith and Jones are of equal mass, which one of the sketches below gives the correct speed-time relationship for the two skaters?

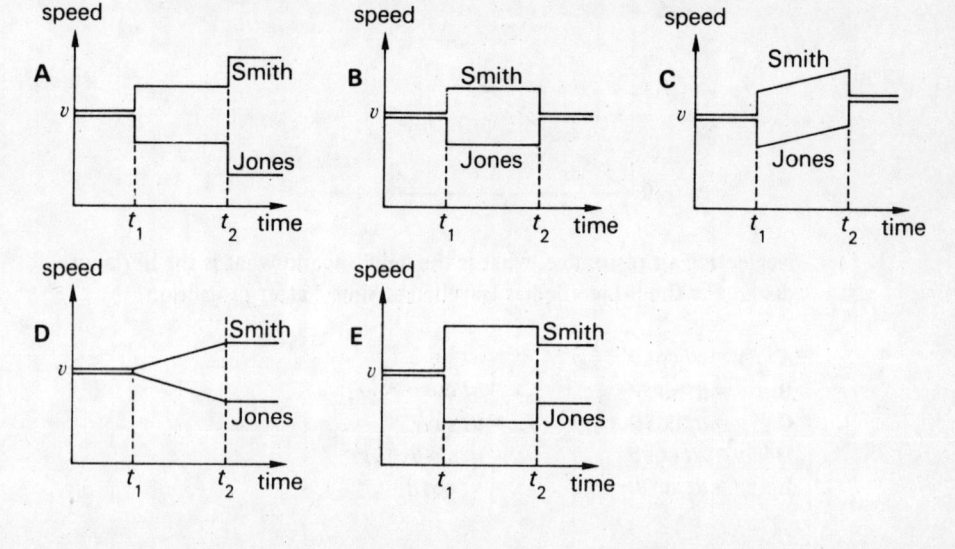

SECTION A   15

A 4.6   The outboard motor of a small boat has a propeller of diameter 200 mm. When the boat is at rest, the propeller sends back a stream of water at a speed of $10\,\mathrm{m\,s^{-1}}$. One half of the work that is being done by the motor is transferred to this water as kinetic energy. What is the power output of the motor?

A   1.25 kW   B   6.50 kW   C   15.7 kW   D   31.4 kW
E   125 kW

[Take the density of water to be $1000\,\mathrm{kg\,m^{-3}}$.]

A 4.7   A particle travels in a circle of radius r with a constant angular velocity ω as illustrated below. The particle is at S at time $t = 0$ and at P at time t. Q represents the projection of point P on to the diameter through S. Measured with respect to the origin O, the displacement, linear velocity and linear acceleration of the point Q along this diameter are y, v and a respectively.

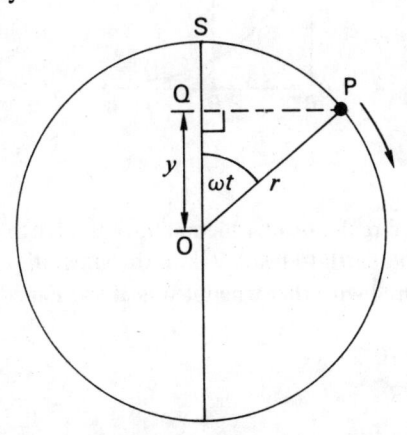

Which one of the following sets of expressions is correct?

A $y = r \cos \omega t$ $v = -r\omega \sin \omega t$ $a = r\omega^2 \cos \omega t$
B $y = r \cos \omega t$ $v = -r\omega \sin \omega t$ $a = -r\omega^2 \cos \omega t$
C $y = r \cos \omega t$ $v = -r\omega \cos \omega t$ $a = -r\omega^2 \sin \omega t$
D $y = r \sin \omega t$ $v = r\omega \cos \omega t$ $a = -r\omega^2 \sin \omega t$
E $y = r \sin \omega t$ $v = r\omega \cos \omega t$ $a = r\omega^2 \sin \omega t$

16 EXERCISE 4

A 4.8 If the Earth were a sphere of radius R and uniform density, which one of the following graphs would represent the force of gravity F on a body at distances x from the centre of the Earth?

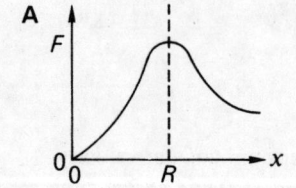

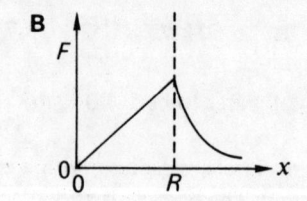

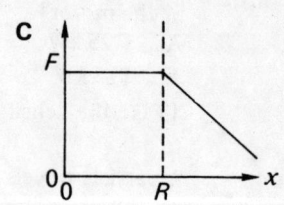

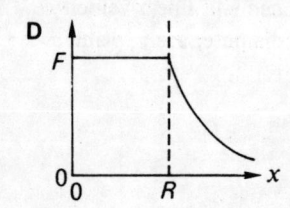

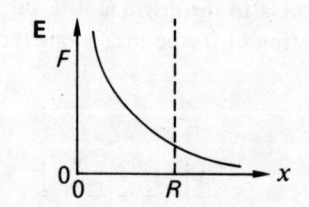

A 4.9 A satellite is in circular orbit 144 km above the Earth's surface. Assuming the radius of the Earth to be 5760 km, the gravitational force on the satellite compared with that when it was at the Earth's surface is (approximately)

    A   greater by 10%.
    B   greater by 5%.
    C   the same.
    D   less by 5%.
    E   less by 10%.

A 4.10 Two immiscible liquids P and Q of different densities are contained in a wide U-tube as shown below. The heights of the two liquids above the horizontal line X X' which cuts the boundary between the liquids are H_P and H_Q respectively.

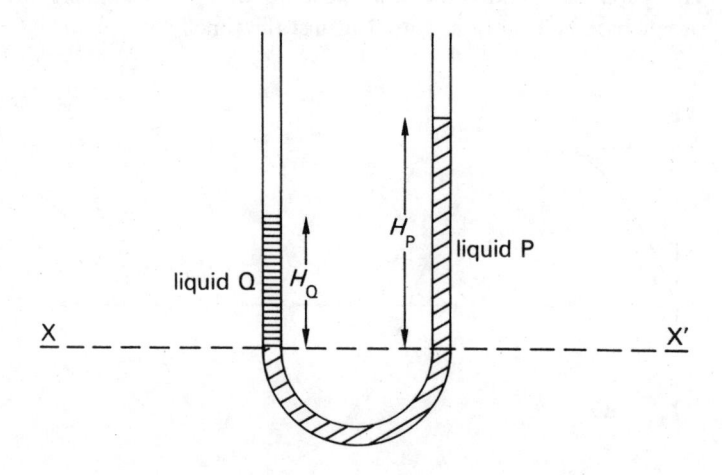

The U-tube is transported to a planet where the acceleration of free fall is 2/3 that on Earth, where the liquid does not evaporate and where the heights of liquid (measured relative to X X') are h_P and h_Q respectively. Which one of the following statements is correct?

A   The liquid levels are unchanged, so that $h_P = H_P$ and $h_Q = H_Q$
B   Both liquid levels rise so that $h_P/h_Q = H_P/H_Q$
C   Both liquid levels rise so that $(h_P - h_Q) = (H_P - H_Q)$
D   Liquid P rises and liquid Q falls so that $h_P/h_Q = 3H_P/2H_Q$
E   Liquid P falls and liquid Q rises so that $h_P/h_Q = 2H_P/3H_Q$

SECTION B OSCILLATIONS AND WAVES

EXERCISE 1

B 1.1   The graph below shows the displacement x of a particle undergoing linear simple harmonic motion plotted against the time t.

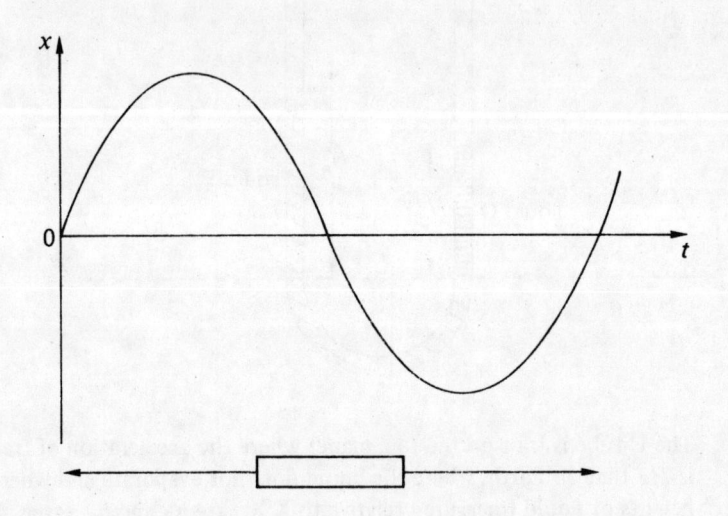

What is the appropriate label for the box in the diagram?
A   amplitude
B   frequency
C   period
D   phase
E   wavelength

B 1.2   A particle executes simple harmonic motion of amplitude 2.0×10^{-3} m and period 0.10 s. Its maximum speed is approximately
A 3.2×10^{-5} m s^{-1}
B 2.0×10^{-4} m s^{-1}
C 2.0×10^{-2} m s^{-1}
D 1.3×10^{-1} m s^{-1}
E 5.0×10^{4} m s^{-1}

B 1.3 Sound travels through air as
- **A** a longitudinal progressive wave.
- **B** a transverse progressive wave.
- **C** a longitudinal stationary wave.
- **D** a transverse stationary wave.
- **E** a combination of longitudinal and transverse waves.

B 1.4 Which of the following statements best describes what occurs when a parallel beam of light initially in one medium strikes the boundary with a denser medium at a large angle to the normal?
- **A** The beam is totally reflected at the boundary.
- **B** The beam is bent away from the normal, because light travels faster in the denser medium.
- **C** The beam is bent towards the normal, because light travels faster in the denser medium.
- **D** The beam is bent away from the normal, because light travels more slowly in the denser medium.
- **E** The beam is bent towards the normal, because light travels more slowly in the denser medium.

B 1.5 Which one of the following statements about two wave-trains of monochromatic light arriving at a point on a screen must be true if the wave-trains are coherent?
- **A** They are in phase.
- **B** They have a constant phase difference.
- **C** They have both travelled paths of equal length.
- **D** They have approximately equal amplitudes.
- **E** They interfere constructively.

B 1.6 Diffraction is the name given to
- **A** the effect observed at the boundary when a beam of light passes from one transparent medium to another.
- **B** the emission of X-rays from crystals.
- **C** the splitting of light into its component wavelengths by means of a prism.
- **D** the superposition of two beams of radiation of the same wavelength and bearing a constant phase relation to one another.
- **E** the effect by which radiation enters the geometrical shadow of an opaque object.

EXERCISE 1

B 1.7 A parallel beam of monochromatic light of wavelength λ is incident normally on a diffraction grating G. The angle between the directions of the second-order spectrum S_1 on one side and the second-order spectrum S_2 on the other side is α, as shown below.

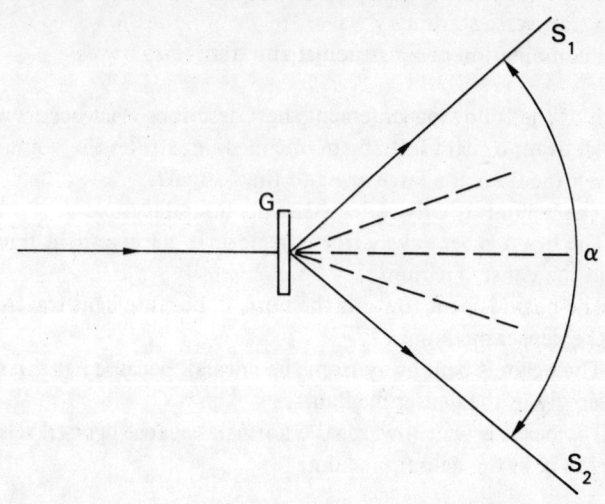

What is the distance between the centres of adjacent lines on the grating?
A $4\lambda/\sin \alpha$ B $2\lambda/\sin \alpha$ C $\lambda/\sin \alpha$ D $2\lambda/\sin \frac{1}{2}\alpha$
E $\lambda/\sin \frac{1}{2}\alpha$

B 1.8 The taut sonometer wire shown below is held momentarily at its centre while being plucked at a distance $l/4$ from one of the bridges.

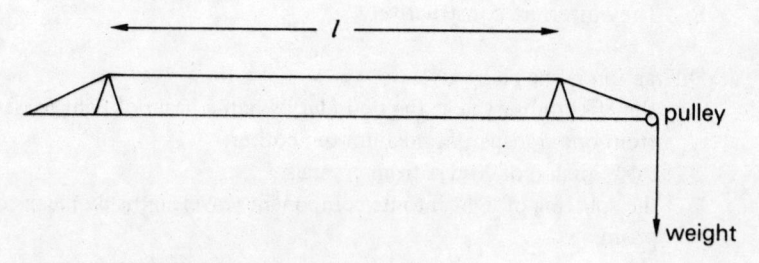

The *predominant* note heard is the
A fundamental.
B second harmonic.
C third harmonic.
D fourth harmonic.
E eighth harmonic.

SECTION B

B 1.9 An air-filled resonance tube, open at both ends and resonating to a tuning fork
    A   always has a central vibration node.
    B   always has a central vibration antinode.
    C   always has an odd number of vibration nodes.
    D   always has an even number of vibration nodes.
    E   always has an odd number of vibration nodes + antinodes.

B 1.10 X-rays, visible light and radio waves are all electromagnetic waves. What is the correct sequence if these are arranged in decreasing order of wavelength?

A	X-rays	visible	radio
B	visible	X-rays	radio
C	visible	radio	X-rays
D	radio	visible	X-rays
E	X-rays	radio	visible

EXERCISE 2

B 2.1 When a particle performs linear simple harmonic motion
    A   its acceleration and speed are zero at the centre of the motion.
    B   its acceleration and speed have their least values at the centre of the motion.
    C   its acceleration and speed have their greatest values at the extremities of the motion.
    D   its acceleration is zero and its speed has its greatest value at the extremities of the motion.
    E   its acceleration has its greatest value and its speed is zero at the extremities of the motion.

B 2.2 When a particle performs simple harmonic motion the velocity leads the displacement by a phase angle of
    A $\pi/4$ rad   B $\pi/2$ rad   C $3\pi/4$ rad   D π rad
    E   zero

B 2.3 Which of the quantities, mass and energy, will be transported in the direction of propagation of a plane sound wave through a medium?
    A   mass only
    B   energy only
    C   both mass and energy
    D   neither mass nor energy
    E   the answer depends on the medium

22 EXERCISE 2

B 2.4 When visible light passes from air into glass, the radiation experiences a change in
- **A** frequency, but not in speed and not in wavelength.
- **B** frequency and speed, but not in wavelength.
- **C** frequency, wavelength and speed.
- **D** wavelength and frequency, but not in speed.
- **E** wavelength and speed, but not in frequency.

B 2.5 A deduction to be made from Young's double-slit experiment is that light
- **A** is electromagnetic in nature.
- **B** consists of transverse oscillations.
- **C** is a wave motion.
- **D** consists of photons.
- **E** travels in air at 3×10^8 m s^{-1}

B 2.6 Which one of the following is a consequence of diffraction alone?
- **A** the splitting of white light into a spectrum
- **B** the deviation of a wave around the edge of an obstacle
- **C** the production of a line spectrum by a grating
- **D** the blurring of a shadow when using an extended source
- **E** the interaction of wavefronts, causing maxima and minima of intensity

B 2.7 A diffraction grating is ruled with 650 lines per millimetre. When monochromatic light falls normally on it, the first-order diffracted beams are observed on the far side making an angle of 30° with the normal. What is the frequency of the light?
- **A** 2.3×10^9 Hz
- **B** 2.2×10^{14} Hz
- **C** 3.1×10^{14} Hz   [Take the speed of light to be 3.0×10^8 m s^{-1}]
- **D** 3.9×10^{14} Hz
- **E** 7.7×10^{14} Hz

B 2.8 In order to double the frequency of the fundamental note emitted by a stretched string, one can reduce the length to $\frac{3}{4}$ of the original length and then change the tension by a factor

A $\frac{3}{8}$ **B** $\frac{2}{3}$ **C** $\frac{8}{9}$ **D** $\frac{16}{9}$ **E** $\frac{9}{4}$

SECTION B

B 2.9 A vibrating tuning fork is held near the open end of a tube closed at the other end by a movable piston, and the piston is adjusted so that the air column in the tube is the shortest possible length for resonance. What is this length, in terms of the wavelength of the note produced?
A slightly less than half a wavelength
B exactly half a wavelength
C slightly more than half a wavelength
D slightly less than a quarter of a wavelength
E exactly a quarter of a wavelength

B 2.10 Which one of the following summarises the change in wave characteristics on going from infrared to X-rays in the electromagnetic spectrum?

	frequency	*wavelength (in vacuum)*	*speed (in vacuum)*
A	remains constant	decreases	decreases
B	decreases	increases	decreases
C	increases	increases	increases
D	decreases	increases	remains constant
E	increases	decreases	remains constant

EXERCISE 3

B 3.1 A body is performing linear simple harmonic motion about a point. Taking x as the instantaneous distance of the body from the centre of its motion and v as its corresponding speed, the acceleration of the body has its greatest value when
A v and x are both zero.
B v is zero and x is a maximum.
C v is a maximum and x is zero.
D v and x are both maxima.
E the motion has lasted for a long time.

B 3.2 A small mass executes linear simple harmonic motion about a point O with amplitude a and period T. Its displacement from O at time $T/8$ after passing through O is

A $a/8$ B $a/2\sqrt{2}$ C $a/2$ D $a/\sqrt{2}$ E $\frac{(2\sqrt{2})a}{3}$

24 EXERCISE 3

B 3.3 The two displacement time graphs below, drawn to exactly the same scale, are sinusoidal and represent the motion of the air in the path of each of two sound waves.

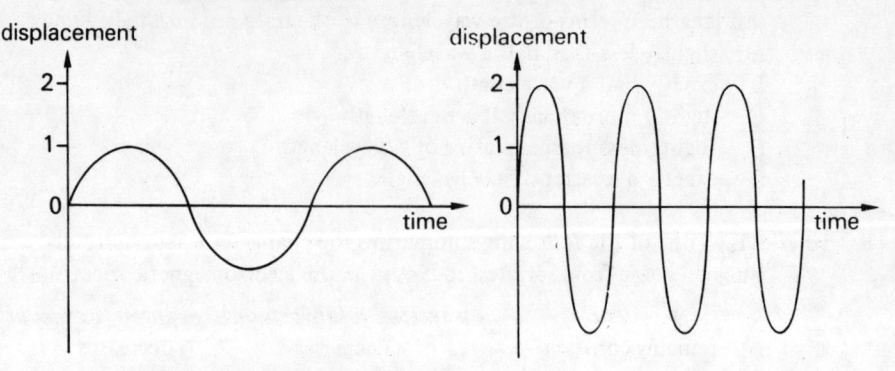

The two sound waves have
- **A** the same pitch but different quality.
- **B** different pitch and quality.
- **C** different intensity and quality.
- **D** the same pitch but different intensity.
- **E** the same quality but different intensity.

B 3.4 Light travelling through a medium of refractive index n_1 has a speed v, wavelength λ, and frequency f. It then enters a second medium of refractive index n_2. Which one of the following describes the characteristics of the wave in the second medium?

	speed	wavelength	frequency
A	$\dfrac{n_1 v}{n_2}$	$\dfrac{n_1 \lambda}{n_2}$	f
B	$\dfrac{n_2 v}{n_1}$	$\dfrac{n_2 \lambda}{n_1}$	f
C	$\dfrac{n_2 v}{n_1}$	λ	$\dfrac{n_2 f}{n_1}$
D	$\dfrac{n_2 v}{n_1}$	λ	$\dfrac{n_1 f}{n_2}$
E	$\dfrac{n_1 v}{n_2}$	$\dfrac{n_1 \lambda}{n_2}$	$\dfrac{n_1 f}{n_2}$

SECTION B

B 3.5 The diagram below represents a typical Young's slits arrangement, in which a is the width of the primary slit P, l its distance from the secondary slits S_1 and S_2, s the separation of the secondary slits, d the distance of the screen from them, and x the separation of adjacent bright bands on the screen.

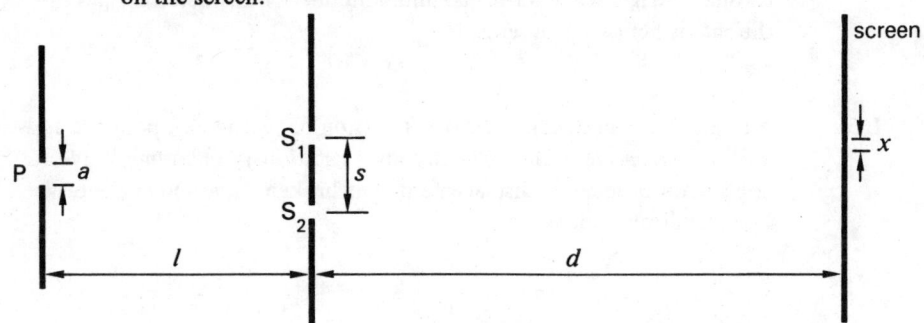

Taking λ as the wavelength of the monochromatic light used, then the separation x is given by

A $\dfrac{\lambda s}{d}$ **B** $\dfrac{\lambda d}{s}$ **C** $\dfrac{\lambda l}{d}$ **D** $\dfrac{\lambda l}{a}$ **E** $\dfrac{\lambda^2 a}{ld}$

B 3.6 A narrow vertical single slit is illuminated by a parallel beam of monochromatic light. Which of the following diagrams represents how the distribution of the emergent diffracted light varies with horizontal direction?

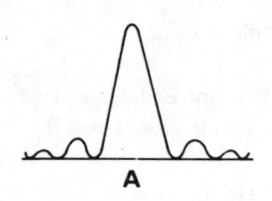

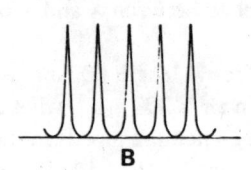

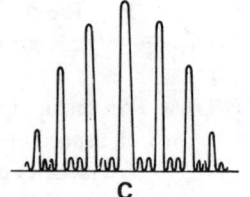

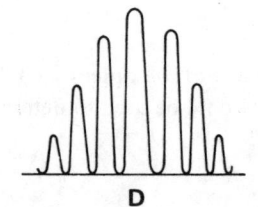

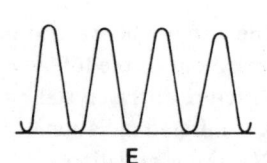

26 EXERCISE 3

B 3.7 Monochromatic light of wavelength 600 nm is used in a spectrometer to illuminate a diffraction grating set normally to the collimator. The grating has 3×10^5 lines per metre. The telescope is used to scan the field to *one side* of the 'straight-through' position. Not counting the 'straight-through' image, what is the maximum number of diffracted images of the slit visible to the observer?

A 2   **B** 5   **C** 8   **D** 10   **E** 11

B 3.8 A string is stretched under constant tension between fixed points X and Y. The solid curve in the diagram shows a stationary (standing) wave at one instant of greatest displacement. The broken curve shows the other extreme displacement.

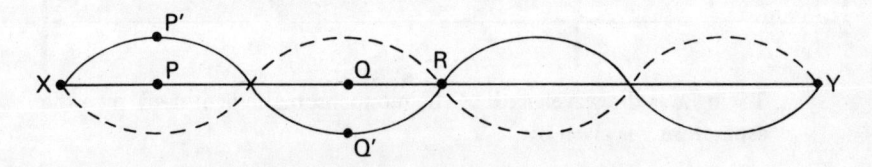

Which one of the following statements is correct?
A The distance between P and Q is one wavelength.
B A short time later, the string at R will be displaced.
C The string at P' and the string at Q' will next move in opposite directions to one another.
D At the moment shown, the energy of the stationary wave is all in the form of kinetic energy.
E The stationary wave shown has the lowest possible frequency for this string stretched between X and Y under this tension.

B 3.9 An organ pipe of effective length 600 mm is closed at one end. Given that the speed of sound in air is 300 m s^{-1}, what are the two lowest resonant frequencies produced when the pipe is sounded?
A 125 Hz and 250 Hz   **B** 125 Hz and 375 Hz
C 250 Hz and 500 Hz   **D** 250 Hz and 750 Hz
E 500 Hz and 1000 Hz

B 3.10 The distance between atoms in a crystal is of the order of 10^{-8} cm. Which one of the following radiations would be used to determine the structure of the crystal by diffraction?
A infrared radiation
B visible radiation
C ultraviolet radiation
D X-rays
E γ-rays

EXERCISE 4

B 4.1 Which one of the following sketch graphs best represents the relation between the acceleration a of a body executing a simple harmonic motion and the displacement x of the body from the centre of its path?

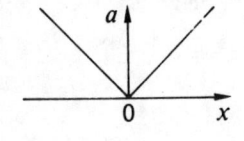

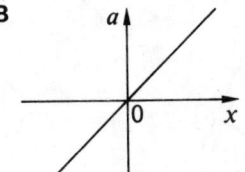

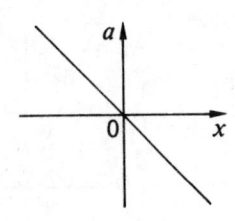

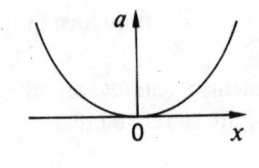

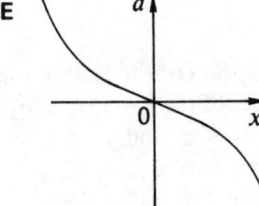

A   B   C

D   E

B 4.2 A particle moves with simple harmonic motion along an x-axis according to the equation

$$\frac{d^2x}{dt^2} + Ax = 0$$

The period of this motion is

A $\dfrac{\sqrt{A}}{2\pi}$ B $\dfrac{\sqrt{A}}{\pi}$ C $\dfrac{\pi}{\sqrt{A}}$ D $\dfrac{2\pi}{\sqrt{A}}$ E $\dfrac{\pi}{2\sqrt{A}}$

B 4.3 A sound wave of frequency 400 Hz is travelling in air at a speed of 320 m s^{-1}. What is the difference in phase between two points on the wave 1 m apart in the direction of travel?

A   zero   B $\frac{4}{5}\pi$ rad   C $\frac{8}{5}\pi$ rad   D $\frac{5}{2}\pi$ rad
E 400π rad

EXERCISE 4

B 4.4 The diagram shows a beam of light incident, at an angle i, on the interface between two transparent media 1 and 2. PQ is a wavefront at time 0, and XY is the same wavefront a little later.

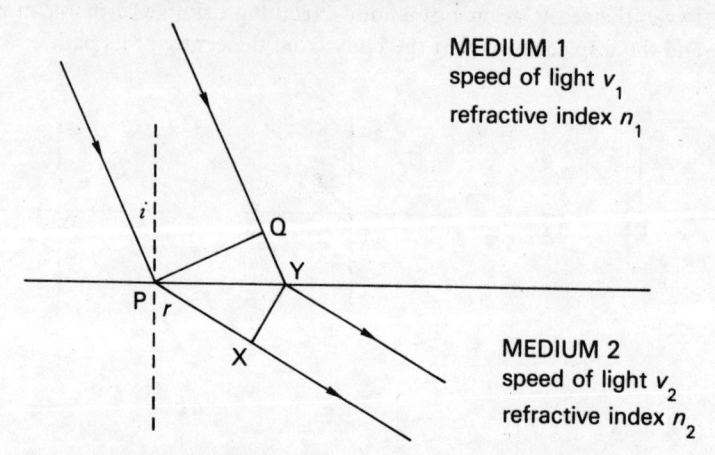

MEDIUM 1
speed of light v_1
refractive index n_1

MEDIUM 2
speed of light v_2
refractive index n_2

The speeds of the light in the two media v_1 and v_2, and the refractive indices of the two media, n_1 and n_2 are so related that

A $v_1 < v_2$ and $n_1 > n_2$.
B $v_1 < v_2$ and $n_1 < n_2$.
C $v_1 > v_2$ and $n_1 > n_2$.
D $v_1 > v_2$ and $n_1 < n_2$.
E $v_1 = v_2$ and $n_1 = n_2$.

B 4.5 The intensity of a wave is proportional to the square of the amplitude of the wave. If two light waves of the same frequency and phase are superimposed, the total intensity is proportional to

A the sum of the intensities of the separate waves.
B the mean value of the intensities of the separate waves.
C the square of the sum of the two amplitudes.
D the square of the difference of the two amplitudes.
E the square of the mean value of the two amplitudes.

B 4.6 When you look at a distant point source of monochromatic light through a piece of gauze with a fine square mesh you see

A a pattern of concentric circles.
B a pattern of parallel lines.
C two lines at right-angles.
D a pattern of separate points.
E a pattern of radial lines.

B 4.7 A parallel beam of white light (range of wavelengths 4.5×10^{-7} m to 7.5×10^{-7} m) is incident normally on a diffraction grating. The most deviated wavelength in the second order spectrum is diffracted through an angle of 60° from the direction of the incident beam. How many lines per metre are there on the grating?
A 5.8×10^5 B 9.6×10^5 C 11.6×10^5
D 19.2×10^5 E 2.3×10^6

B 4.8 A suspension bridge is to be built across a valley where it is known that wind gusts at regular 5 s intervals can occur. It is estimated that the speed of transverse waves along the span of the bridge would be 400 m s^{-1}. It is possible that, with such wind gusts, the bridge might resonate dangerously at its fundamental frequency if the span had a length of
A 2000 m   B 1000 m   C 400 m   D 80 m
E 40 m

B 4.9 Air-filled pipes of equal length are represented in the diagrams below.

closed   closed   closed   open   open   open

fundamental f_a f_b f_c
frequency

In what ratio are the frequencies of the *fundamental* vibrations of the air columns in the pipes?
$f_a : f_b : f_c$
A $1 : 2 : 1$
B $1 : 2 : 2$
C $1 : \frac{1}{2} : \frac{1}{4}$
D $1 : \frac{1}{2} : 1$
E $1 : \frac{1}{2} : 2$

B 4.10 What is the approximate range of wavelengths of infrared radiation?
A 10^{-9} m to 10^{-7} m
B 10^{-7} m to 10^{-6} m
C 10^{-6} m to 10^{-3} m
D 10^{-4} m to 10^{+4} m
E 10^{-1} m to 10^{+2} m

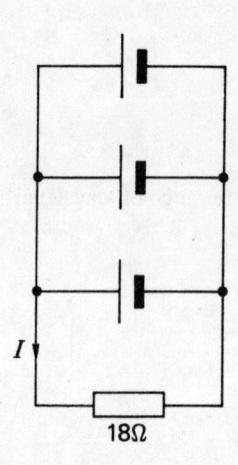

SECTION C ELECTRICITY (i): D.C. CIRCUITS AND ELECTROSTATICS

EXERCISE 1

C 1.1 If two wires of resistance R_1 and R_2 are connected together in parallel, the effective resistance is

  A $R_1 + R_2$ B $\frac{1}{2}(R_1 + R_2)$ C $(R_1^2 + R_2^2)^{\frac{1}{2}}$

  D $\dfrac{R_1 R_2}{R_1 - R_2}$ E $\dfrac{R_1 R_2}{R_1 + R_2}$

C 1.2 A low-resistance milliammeter may be converted into a voltmeter by the connection of
  A a high resistance in series.
  B a high resistance in parallel.
  C a low resistance in series.
  D a low resistance in parallel.
  E a high resistance in series and a low resistance in parallel.

C 1.3 A resistor and three similar cells, each of e.m.f. 1.5 V and internal resistance 6.0 Ω, are connected as shown below.
  The current I through the resistor is
  A 0.075 A
  B 0.083 A
  C 0.24 A
  D 0.36 A
  E 0.45 A

SECTION C

C 1.4 If P is the power dissipated in a resistance maintained at constant temperature and V is the potential difference across the resistance,
- A P is proportional to V^{-2}
- B P is proportional to V^{-1}
- C P is proportional to $V^{\frac{1}{2}}$
- D P is proportional to V
- E P is proportional to V^2

C 1.5 What is the electrical capacitance of a conductor?
- A the charge stored in unit volume of the conductor
- B the charge stored on unit area of the conductor
- C the work done in bringing unit charge up from infinity to the conductor
- D the charge required to raise the potential of the conductor by one unit
- E the potential required to store unit charge on the conductor

C 1.6 Four capacitors are connected as shown below.

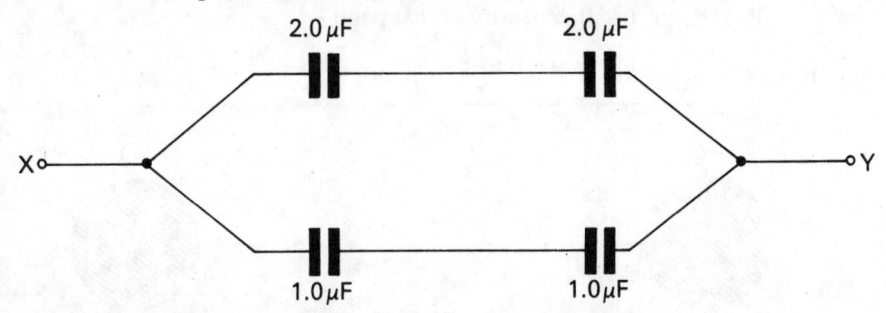

What is the effective capacitance of the combination between X and Y?
- A $1.3\,\mu F$
- B $1.5\,\mu F$
- C $3.0\,\mu F$
- D $4.5\,\mu F$
- E $6.0\,\mu F$

C 1.7 The energy stored in a capacitor which carries charge Q at potential difference V is $\frac{1}{2}QV$. Why does the factor $\frac{1}{2}$ occur in this formula?
- A The charge Q is stored half on one plate and half on the other.
- B The average potential difference against which the charge is put on the plates is $\frac{1}{2}V$.
- C When a capacitor is charged half the charge is lost to the surroundings.
- D Half the energy is stored in the dielectric when the capacitor is charged.
- E Half the energy is dissipated as heat in the dielectric when the capacitor is charged.

EXERCISE 1

C 1.8 A capacitor is charged from a 4 V supply 50 times per second and discharged through a milliammeter 50 times per second by means of a vibrating switch. If the milliammeter reads 1 mA, what is the value of the capacitance?
- A $1\,\mu F$
- B $2\,\mu F$
- C $5\,\mu F$
- D $8\,\mu F$
- E $10\,\mu F$

C 1.9 What is the electric potential at a point in an electric field?
- A   the force per unit charge acting on a small positive charge at that point
- B   the charge density at that point
- C   the work done in bringing unit positive charge from infinity to that point
- D   the force exerted on a wire carrying unit current and passing through that point
- E   the electric flux density at that point

C 1.10

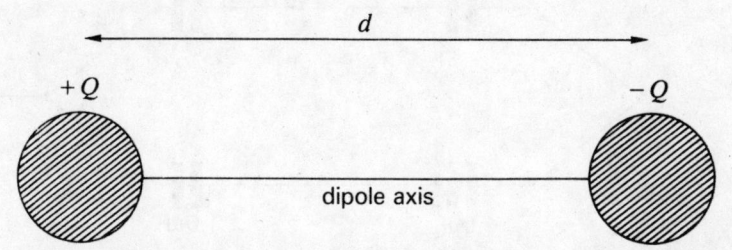

The figure shows a stationary electric dipole placed in a uniform electric field of intensity E with its axis perpendicular to the field. The dipole experiences
- A   no resultant force or couple.
- B   a resultant force of $2QE$ and no couple.
- C   no resultant force and a couple of moment QdE.
- D   a resultant force $2QE$ and a couple of moment QdE.
- E   a resultant force $2QE$ and a couple of moment $2QdE$.

EXERCISE 2

C 2.1 The circuit shown below contains a 6 V battery of negligible internal resistance.

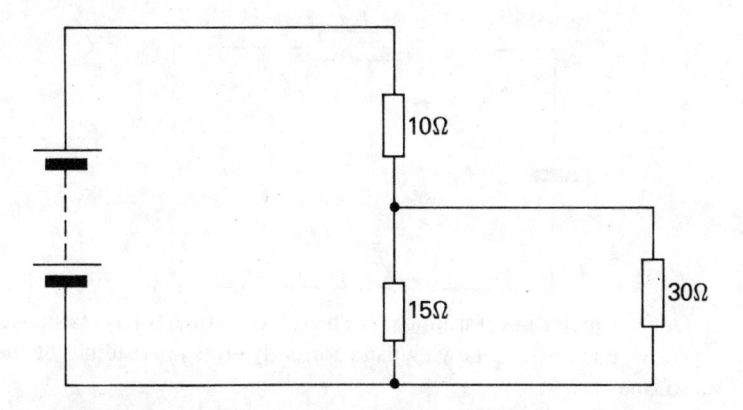

What is the current flowing in the 30 Ω resistor?

A   0.1 A   B   0.2 A   C   0.3 A   D   0.4 A   E   0.5 A

C 2.2 A moving-coil meter of resistance 100 Ω gives its full-scale deflection when a current of 100 μA is passed through it. If this meter is intended to give a full-scale deflection when measuring a current of 1.00 mA it must be shunted by a resistor of value

A   9.00 Ω   B   10.0 Ω   C   11.1 Ω   D   900 Ω
E   1 110 Ω

34 EXERCISE 2

C 2.3 A battery of e.m.f. E and negligible internal resistance is connected as shown to a resistor of resistance R.
Two voltmeters are used, one at a time, to measure the potential difference across the terminals X and Y.

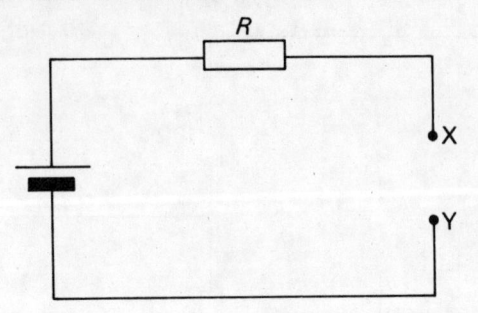

One voltmeter has resistance R. The other voltmeter has resistance $10R$. Which one of the following pairs correctly gives the readings of the two voltmeters?

	reading of voltmeter of resistance R	reading of voltmeter of resistance $10R$
A	nearly E	nearly E
B	$\frac{1}{2}E$	nearly E
C	nearly E	$\frac{1}{2}E$
D	$\frac{1}{2}E$	$\frac{1}{2}E$
E	$\frac{1}{10}E$	nearly E

C 2.4 A resistor of resistance $1.0\,\text{k}\Omega$ has a thermal capacity of $5.0\,\text{J K}^{-1}$. A potential difference of $4.0\,\text{V}$ is applied across it for $120\,\text{s}$. If the resistor is thermally insulated, what is the final rise in temperature?
A $8.0 \times 10^{-4}\,\text{K}$ B $3.2 \times 10^{-3}\,\text{K}$ C $9.6 \times 10^{-2}\,\text{K}$
D $0.38\,\text{K}$ E $9.6\,\text{K}$

C 2.5 A $0.20\,\mu\text{F}$ capacitor is charged to a potential difference of $4.0\,\text{V}$ and then discharged through a ballistic galvanometer (whose deflection is proportional to the charge passing through it). A reading of 24 divisions is recorded. What deflection will occur when a $0.10\,\mu\text{F}$ capacitor charged to $6.0\,\text{V}$ is discharged through the galvanometer?
A 8 divisions
B 12 divisions
C 18 divisions
D 36 divisions
E 72 divisions

SECTION C

C 2.6   A $2\,\mu F$ capacitor is charged to a potential difference of 200 V and then isolated. When it is connected across a second uncharged capacitor, the common potential difference becomes 40 V. What is the capacitance of the second capacitor?

       A $2\,\mu F$ B $4\,\mu F$ C $6\,\mu F$ D $8\,\mu F$ E $16\,\mu F$

C 2.7   A capacitor of capacitance $160\,\mu F$ is charged to a potential difference of 200 V and then connected across a discharge tube, which conducts until the potential difference across it has fallen to 100 V. The energy dissipated in the tube is

       A   6.4 J   B   4.8 J   C   3.2 J   D   2.4 J   E   0.8 J

C 2.8   A circuit consists of a $2\,\mu F$ capacitor (initially uncharged), a $25\,k\Omega$ resistor, a switch and a 100 V battery connected in series. When the switch is closed the initial current is

       A   zero   B   4 mA   C   40 mA   D   2 A   E   4 A

C 2.9   A charge of 3 C is moved from infinity to a point X in an electric field. The work done in this process is 15 J. The electric potential at X is

       A   45 V   B   22.5 V   C   15 V   D   5 V   E   0.2 V

C 2.10   The electric potentials V are measured at distances x from P along a line PQ. The results are

V	13 V	15 V	18 V	21 V	23 V
x	0.02 m	0.03 m	0.04 m	0.05 m	0.06 m

The component along PQ of the electric field strength for $x = 0.04\,m$ is approximately

A $75\,V\,m^{-1}$ towards P.
B $300\,V\,m^{-1}$ towards Q.
C $300\,V\,m^{-1}$ towards P.
D $450\,V\,m^{-1}$ towards Q.
E $450\,V\,m^{-1}$ towards P.

EXERCISE 3

C 3.1 The meter in the circuit shown below has an uncalibrated linear scale. With the circuit as shown, the scale reading is 20.

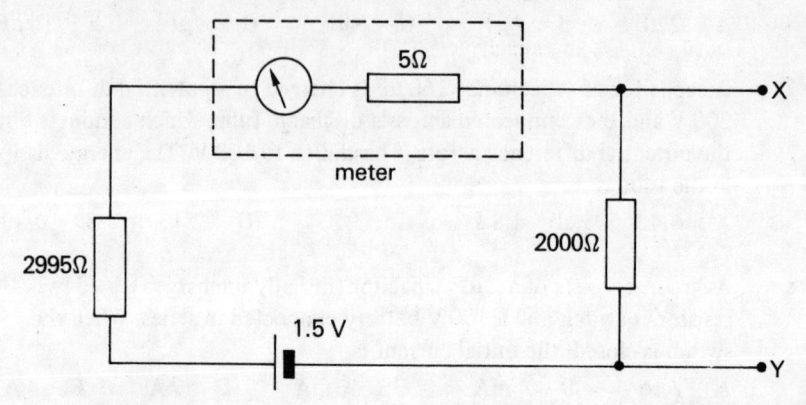

When another 2000 Ω resistor is connected across XY, the scale reading is

A 10   B 16   C 25   D 28   E 40

C 3.2 A galvanometer of resistance 3 Ω shows a full-scale deflection for a current of 10 mA. To adapt this galvanometer for measuring voltages of up to 3 V, it should be used with an external resistance of
A 30 Ω in series.
B 30 Ω in parallel.
C 297 Ω in series.
D 297 Ω in parallel.
E 2997 Ω in series.

C 3.3 A high-resistance voltmeter records the potential difference V across a battery (e.m.f. E and internal resistance r) which is connected to a variable resistance R, as shown below.

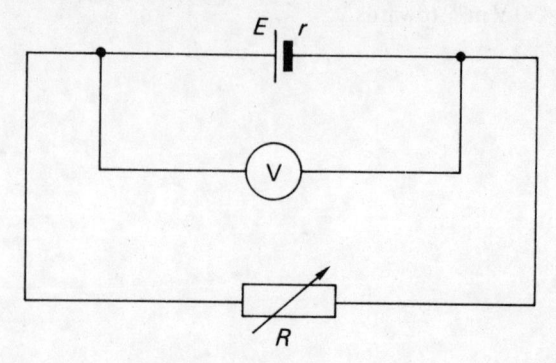

If $V = 6$ V when $R = 2\,\Omega$, and $V = 8$ V when $R = 4\,\Omega$, what is the value of the internal resistance r of the battery?

A $1\,\Omega$ B $2\,\Omega$ C $4\,\Omega$ D $6\,\Omega$ E $8\,\Omega$

C 3.4 An X-ray tube operates at 40 kV with an anode current of 4 mA. The anode has a heat capacity of $20\,\text{J K}^{-1}$. If the anode temperature is to remain constant and the energy radiated is negligible, the rate at which energy must be removed is

A 10 W   B 80 W   C 160 W   D 640 W
E 3200 W

C 3.5 A parallel-plate air-spaced capacitor is charged and isolated. Which one of the following quantities is increased when the plates are moved further apart?
A the capacitance
B the charge on the plates
C the potential difference between the plates
D the electric field strength between the plates
E the force of attraction between the plates

C 3.6 Capacitors C_1 (10 µF) and C_2 (20 µF) are connected in series across a 3 kV supply, as shown.

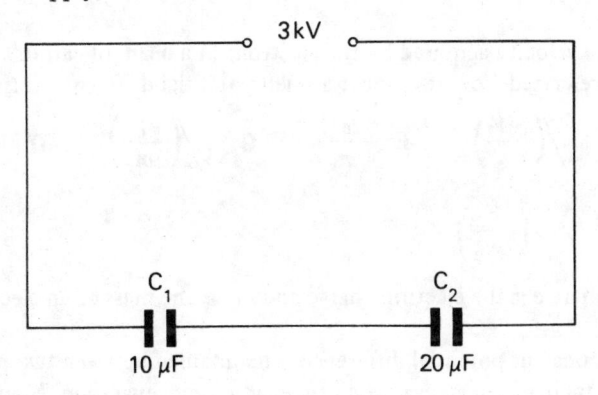

What is the charge on the capacitor C_1?
A 4.5 mC   B 10 mC   C 15 mC   D 20 mC
E 25 mC

C 3.7 A photographic flash unit consists of a xenon-filled flash tube energised by the discharge of a capacitor, previously charged by a 1000 V source. The average power delivered to the tube is 2 000 W, in a time of 0.04 s. The capacitance of the capacitor can be estimated as
A 40×10^{-6} F   B 80×10^{-6} F   C 160×10^{-6} F
D 80×10^{-3} F   E 160×10^{-3} F

EXERCISE 3

C 3.8 A circuit is set up as shown below with the capacitor C initially uncharged.

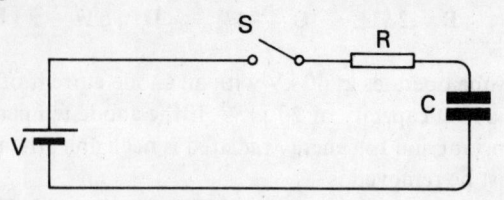

Which one of the following graphs represents the variation with time of the current I in the circuit when the switch S is closed?

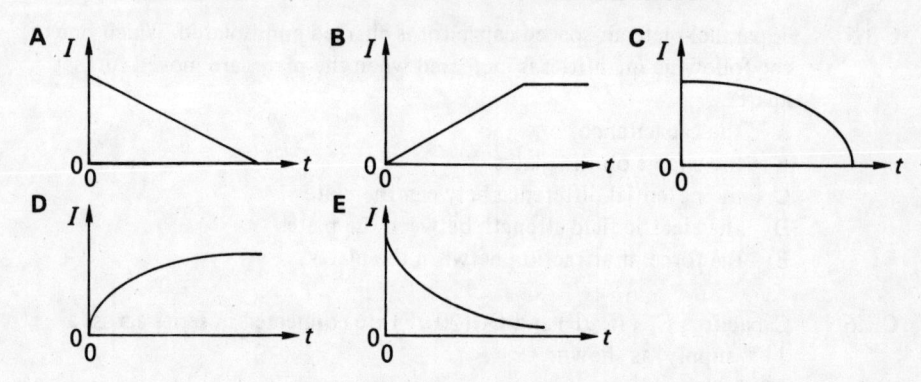

A — I decreases linearly from initial value to 0

B — I increases linearly then levels off

C — I decreases as a concave curve to 0

D — I rises and levels off (saturating)

E — I decays exponentially from initial value to 0

C 3.9 The velocity acquired by the electrons in a beam of cathode rays that are accelerated from rest under a small potential difference V is

A $\sqrt{\left(\dfrac{2eV}{m_e}\right)}$ B $\dfrac{2eV}{m_e}$ C $\sqrt{\left(\dfrac{eV}{2m_e}\right)}$ D $\dfrac{eV}{2m_e}$

E $V\sqrt{\left(\dfrac{e}{2m_e}\right)}$

[where e is the electron charge and m_e is the mass of an electron]

C 3.10 A constant potential difference is maintained between two parallel metal plates in an evacuated tube, and their separation d can be varied. The force experienced by an electron in the space between the plates is proportional to

A d^{-2} B d^{-1} C $d^{\frac{1}{2}}$ D d E d^2

SECTION C

EXERCISE 4

C 4.1 Six resistors each of 5 Ω are connected to form a closed loop (as shown below) and a 10 V battery of negligible internal resistance is connected between P and S.

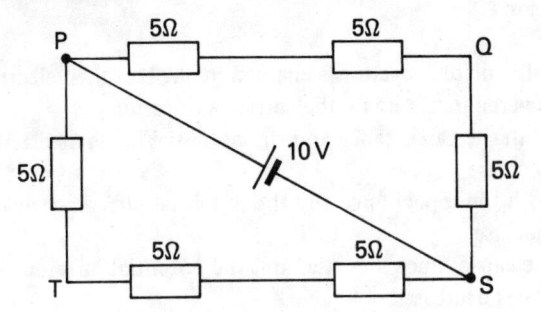

What is the potential difference between Q and T?

A $\frac{10}{3}$ V B $\frac{20}{3}$ V C $\frac{40}{9}$ V D 10 V E $\frac{120}{9}$ V

C 4.2 The diagram below shows the wiring diagram for a multimeter. The different current and voltage ranges are selected by moving the two-pole rotary switch S. The meter M has a resistance of 50 Ω and gives a full-scale deflection when a current of 1 mA passes through it.
At which pair of contacts must S be set for the range 0 V to 1 V?

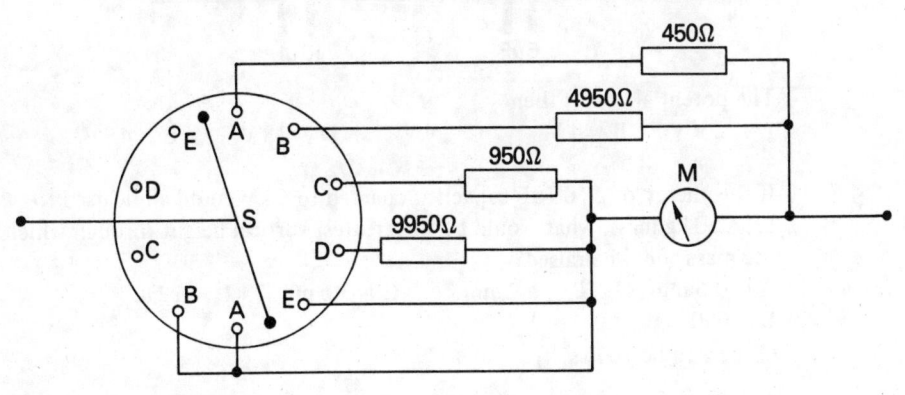

C 4.3 A battery has an electromotive force of 60 V and an internal resistance of 20 Ω. Its terminals are joined by a resistance wire the length of which is varied until the rate of heat production in the wire reaches a maximum value. When this happens, what is the potential difference across the terminals of the battery?

A 60 V B 40 V C 30 V D 20 V E zero

EXERCISE 4

C 4.4 A 40 W heater is to be constructed from wire of radius 0.2 mm and resistivity 2.4×10^{-8} Ω m and is to be used in conjunction with a 12 V car battery. What length of wire is required?

A $\frac{20}{3}\pi$ m B 6π m C $0.6\pi^2$ m D 0.6π m
E $\frac{5}{9}\pi$ m

C 4.5 A parallel-plate capacitor is charged in air. It is then electrically isolated and lowered into a liquid dielectric. As a result,
A both the capacitance and the potential difference across the plates decrease.
B both the capacitance and the potential difference across the plates increase.
C the capacitance increases and the potential difference across the plates decreases.
D both the capacitance and the charges on the plates decrease.
E both the capacitance and the charges on the plates increase.

C 4.6 The diagram below shows two capacitors C_1 (5 μF) and C_2 (10 μF) in series, with point O earthed. The potential of P is raised so that C_1 is given a charge of 30 μC.

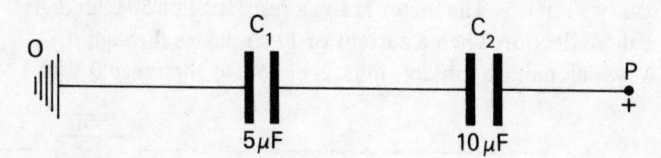

The potential of P is then
A 2 V   B 4 V   C 9 V   D 18 V   E 45 V

C 4.7 If the energy of a 100 μF capacitor charged to 6 kV could all be used to lift a 50 kg mass, what would be the greatest vertical height through which the mass could be raised?
A 0.6 mm   B 1.2 mm   C 3.6 m   D 12 m
E 600 m
[Take g to be 10 m s^{-2}.]

SECTION C

C 4.8   A charged $100\,\mu F$ capacitor is discharged through a $10\,k\Omega$ resistor for 1 second. The ratio

$$\frac{\text{charge remaining on the capacitor}}{\text{original charge on the capacitor}}$$

is

A   0.5   B $ln 2$ C $(1 - ln\,2)$ D $1/e$ E $(1 - 1/e)$

C 4.9   In a certain particle accelerator, doubly-ionised helium atoms (He^{2+}) pass between points which differ in potential by 1.0×10^6 V. The charge on an electron is -1.6×10^{-19} C. The change of energy of each ion is

A 0.4×10^{-13} J   B 0.8×10^{-13} J   C 1.6×10^{-13} J
D 3.2×10^{-13} J   E 6.4×10^{-13} J

C 4.10   Two large horizontal metal plates are mounted parallel to one another and 0.1 m apart in vacuo, and a potential difference of 5 kV is maintained between them. A small charged polystyrene ball of mass 10^{-4} kg is just supported in the electric field between the plates. What is the charge on the ball?

A 5×10^8 C   B 5×10^7 C   C 2×10^{-6} C
D 2×10^{-8} C   E 2×10^{-9} C

[Take g to be $10\,m\,s^{-2}$.]

SECTION D ELECTRICITY (ii): ELECTROMAGNETISM, ALTERNATING CURRENT, CHARGED PARTICLES

EXERCISE 1

D 1.1   In the diagram below the uniformly wound solenoid, of length l, carries an alternating current of constant amplitude. A search coil is placed in different positions along the solenoid.

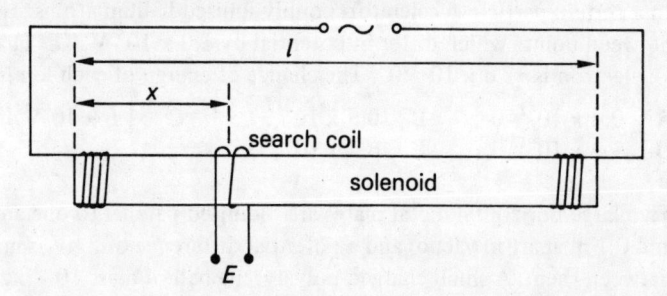

Which one of the following graphs most nearly shows how the amplitude of the e.m.f. E induced in the search coil varies with its position?

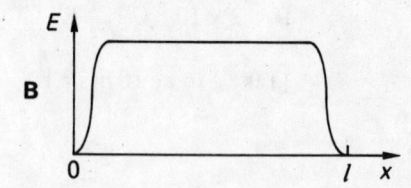

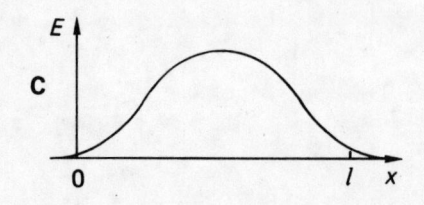

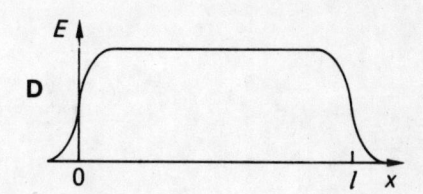

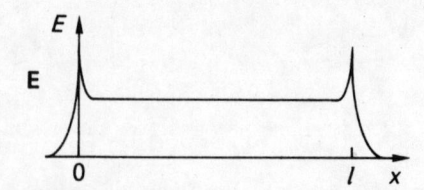

D 1.2 The diagram below shows a wire carrying a current I in the plane of the paper and in the direction shown (north). A magnetic field is applied perpendicular to the paper and acting into the paper.

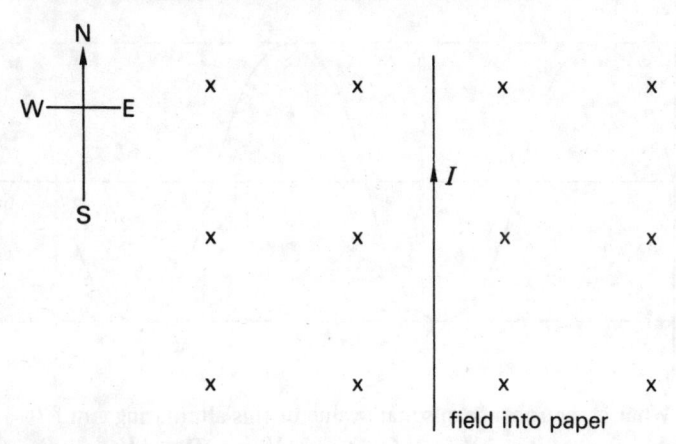

What is the direction of the force acting on the wire?
A   west
B   east
C   south
D   perpendicular to, and into the paper
E   perpendicular to, and out of the paper

D 1.3 The diagram below represents two coils X and Y wound on a soft-iron core.

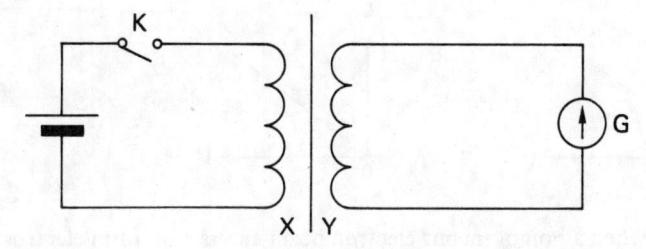

A deflection of the centre-reading galvanometer G is observed
A   momentarily on closing K, but not on opening K again.
B   not on closing K, but momentarily on opening K again.
C   on closing K, and for as long as K remains closed.
D   neither on closing K nor on opening K again.
E   momentarily on closing K, and momentarily on opening K again.

D 1.4 What is the e.m.f. induced in a coil of self inductance 2.0 H, when the current through it is increased uniformly from 10 A to 40 A in 0.15 s?
A   2.0 V   B   5.0 V   C   50 V   D   400 V
E   600 V

44 EXERCISE 1

D 1.5 The graph below represents a sinusoidal alternating e.m.f. of peak value 4 V.

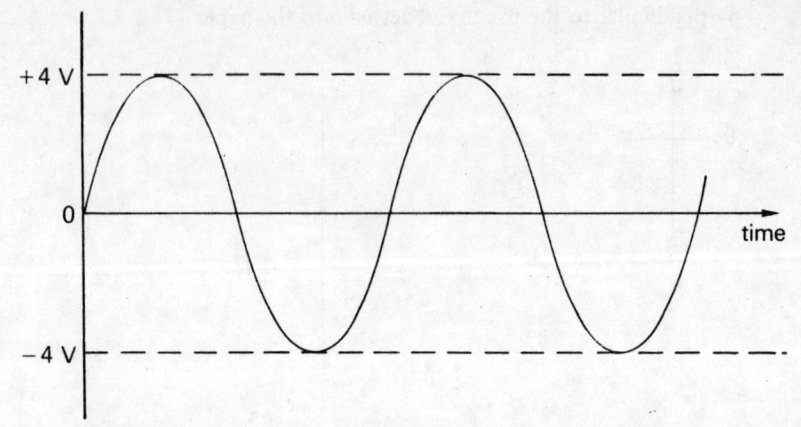

What is the root-mean-square value of this alternating e.m.f.?

   A 0   **B** 2 V   **C** $(2\sqrt{2})$ V   **D** 4 V   **E** $(4\sqrt{2})$ V

D 1.6 Which one of the following graphs best represents the relation between current I and applied potential difference V for the forward characteristic of a germanium diode?

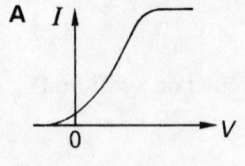

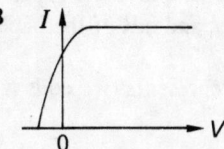

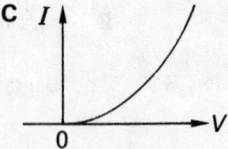

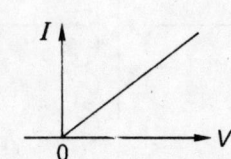

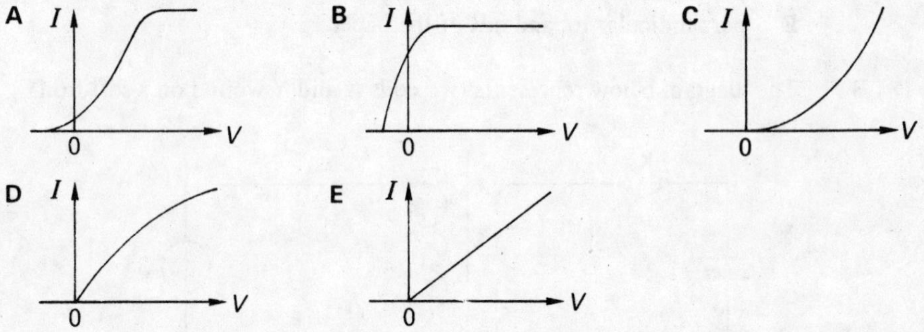

D 1.7 When a homogeneous electron beam enters a uniform electrostatic field which is at right angles to the original direction of the beam, its path in the field is

   A a straight line.
   B an arc of a circle.
   C part of a helix.
   D part of a parabola.
   E none of the above.

D 1.8 When an electron moving with constant speed in a vacuum enters a magnetic field in a direction at right angles to the field, its subsequent path is

A   a straight line parallel to the field.
B   part of a parabola in a plane normal to the field.
C   part of a circle in a plane normal to the field.
D   undeviated.
E   a spiral with the direction of the field as axis.

D 1.9   Starting from rest, a proton and an α-particle are accelerated through the same potential difference. What is the ratio of their final speeds, v_p/v_α?
A   1/2   B $1/\sqrt{2}$ C   1/1   D $\sqrt{2}/1$ E   2/1

D 1.10   In which one of the following are the principal charge carriers *not* mobile electrons?
A   a solenoid
B   a carbon resistor
C   a rod of n-type semi-conductor
D   an electrolyte
E   an oscilloscope tube

EXERCISE 2

D 2.1   A rod is placed inside a solenoid through which a direct current I can be passed, and the magnetic flux density B at a point on the axis of the solenoid outside the rod is measured. As the current is increased from zero up to I_1, the curve in the graph marked (1) is followed, but as it is then reduced from I_1 to zero, curve (2) results.

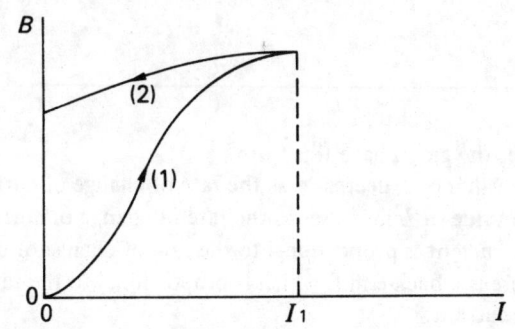

The best explanation of the curves is that
A   the rod is made of a ferromagnetic material, and was magnetised initially.
B   the rod is made of a ferromagnetic material, and remains magnetised when the current returns to zero.
C   there are circulating currents in the rod which produce a magnetic field.
D   there is a back e.m.f. which opposes the changes of the current.
E   the rod is made of a perfectly non-magnetic material.

46 EXERCISE 2

D 2.2 A straight wire 2.0 m long lies at 30° to a uniform magnetic field of flux density 2.0×10^{-5} T and carries a current of 20 mA. What is the magnitude of the force experienced by the wire?

A 10×10^{-8} N B 20×10^{-8} N C 34×10^{-8} N
D 40×10^{-8} N E 67×10^{-8} N

D 2.3 A rectangular coil is rotated with uniform angular velocity about an axis in its own plane at right angles to a uniform magnetic field. At any instant the flux linking the coil is ϕ and the induced e.m.f. is V. When the plane of the coil is at right angles to the field

A both ϕ and V have their maximum values.
B ϕ is a maximum while V is a minimum.
C ϕ is a minimum while V is a maximum.
D both ϕ and V have their minimum values.
E neither ϕ nor V has maximum or minimum values.

D 2.4 A steady potential difference V is applied across a solenoid of self-inductance L and resistance R. When the circuit is first completed, the variation of current I with time t is as shown in the graph below.

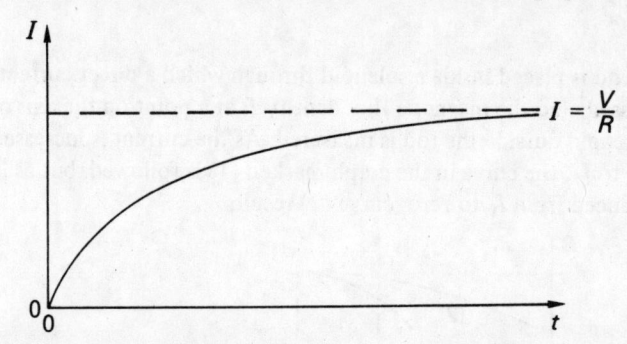

Why does the graph have this form?

A The value of L decreases as the rate of change of current decreases.
B The value of L increases as the rate of change of current decreases.
C The current is proportional to the rate of change of back e.m.f.
D There is a back e.m.f. which is proportional to the rate of change of the current.
E There is a back e.m.f. which is proportional to the current.

D 2.5 An alternating current of root-mean-square value 2 A flowing through a given resistor generates heat at the same rate as a steady direct current I flowing through another resistor of the same value. What is the value of I?

A 1 A B $\sqrt{2}$ A C 2 A D $2\sqrt{2}$ A E 4 A

SECTION D

D 2.6 A p-n junction diode with the forward characteristic shown below is connected in series with a variable low voltage d.c. supply and a 50 Ω resistor.

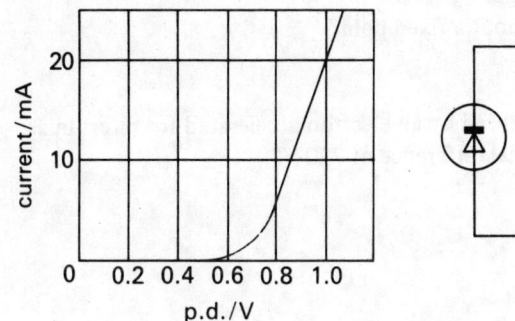

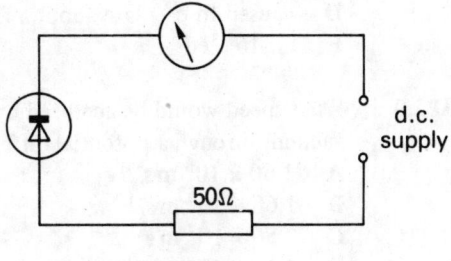

When the meter (which has a negligible resistance) reads 5 mA what is the potential difference across the supply?

A 0.25 V   **B** 0.75 V   **C** 1.05 V   **D** 1.25 V
E 2.75 V

D 2.7 The diagram below shows the trace obtained on an oscilloscope screen when a sinusoidal voltage of frequency f is applied to the X-plates instead of the time base and a second sinusoidal voltage is applied to the Y-plates.

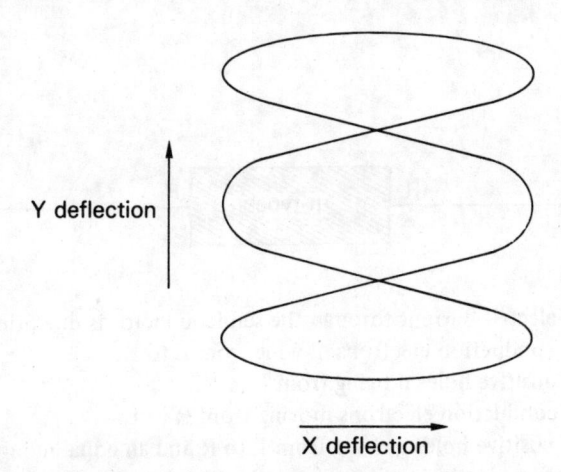

What is the frequency of the voltage applied to the Y-plates?

A $6f$ **B** $3f/2$ **C** $2f/3$ **D** $f/3$ **E** $f/6$

48 EXERCISE 2

D 2.8 If a stationary electron is in a uniform magnetic field it will be
A accelerated in the direction of the field.
B caused to move in a circular path.
C caused to move in an elliptical path.
D caused to oscillate about a fixed point.
E unaffected.

D 2.9 What speed would be acquired by an electron accelerated from rest in a vacuum through a potential difference of 720 V?
A 1.60×10^6 m s^{-1}
B 1.60×10^7 m s^{-1}
C 2.56×10^7 m s^{-1}
D 1.60×10^8 m s^{-1}
E 2.56×10^8 m s^{-1}

[Take the specific charge of the electron e/m_e as 1.78×10^{11} C kg^{-1}.]

D 2.10 A specimen of an n-type semiconductor is connected to a battery and a current I flows, as illustrated in the diagram below.

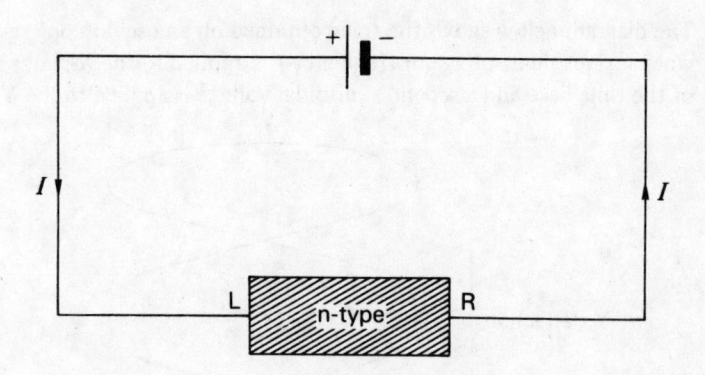

The electric current through the semiconductor is due principally to
A conduction electrons moving from L to R.
B positive holes moving from L to R.
C conduction electrons moving from R to L.
D positive holes moving from L to R and an equal number of conduction electrons moving from R to L.
E positive holes moving from R to L and an equal number of conduction electrons moving from L to R.

EXERCISE 3

D 3.1 Two identical long solenoids, carrying equal currents, are very far apart and the flux density at the end of each is B. The solenoids are now brought into contact co-axially so that the current in each is circulating in the same sense. The flux density on the axis where the solenoids meet is

A zero   B $\frac{1}{2}B$ C B D $\sqrt{2}B$ E $2B$

D 3.2 Two long parallel wires X and Y carry currents of 3 A and 5 A respectively. The force per unit length experienced by X is 5×10^{-5} N to the right as shown in the diagram below.

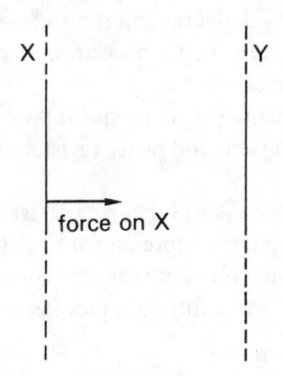

What is the force per unit length experienced by wire Y?
A 2×10^{-5} N to the left
B 3×10^{-5} N to the right
C 3×10^{-5} N to the left
D 5×10^{-5} N to the right
E 5×10^{-5} N to the left

D 3.3 The vertical aerial for a car radio is 1.50 m long. The car is travelling at 20 m s^{-1} on a horizontal road in a (magnetic) east-west direction, at a place where the horizontal component of the Earth's magnetic flux density is 5.0×10^{-6} T. What is the induced e.m.f. between the ends of the aerial?
A zero   B $3.75 \, \mu\text{V}$ C $75.0 \, \mu\text{V}$ D $1000 \, \mu\text{V}$
E $1500 \, \mu\text{V}$

D 3.4 A 1.0 Ω resistor is connected in series with a switch, and an inductor and ammeter both of negligible resistance, across a 12 V d.c. supply. At a certain instant after the switch is closed, the current is 10 A and is increasing at the rate of 500 A s^{-1}. What is the self inductance of the inductor?
A 1.6 mH   B 4.0 mH   C 28 mH   D 60 mH
E 240 mH

EXERCISE 3

D 3.5 One type of ammeter uses the heating effect of a current to produce the deflection of the pointer. The reading, which is proportional to the rate of heating, is X when a direct current I flows through the meter. When inserted in a circuit in which an alternating current of root-mean-square value I flows, the reading is

- **A** $X/2$, because the constantly changing current produces a constantly changing heating effect which averages to one half that of the direct current.
- **B** $X/\sqrt{2}$, because the meter measures the root-mean-square current which is obtained by recalibrating the scale for a.c. use by dividing all scale readings by $\sqrt{2}$.
- **C** X, because the meter measures the root-mean-square current which gives the same deflection on the scale as the direct current.
- **D** $\sqrt{2}X$, because the meter measures the peak current which is $\sqrt{2}$ times the direct current.
- **E** zero, because the needle cannot follow the fast oscillations of the alternating current and hence registers zero on the scale.

D 3.6 In each of the graphs A to E below, voltage V is plotted as a function of time t. Which graph best represents the relationship between voltage and time for the output from a power unit consisting of an alternating-current generator and a full-wave rectifier without smoothing?

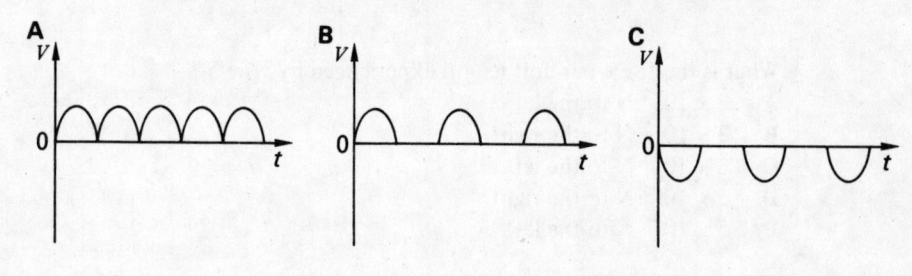

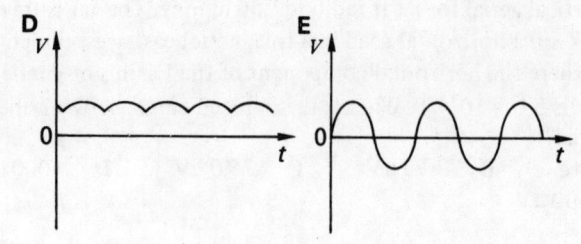

D 3.7 The voltages V_X and V_Y shown below are applied to the X- and Y-plates respectively of a cathode-ray oscilloscope tube.

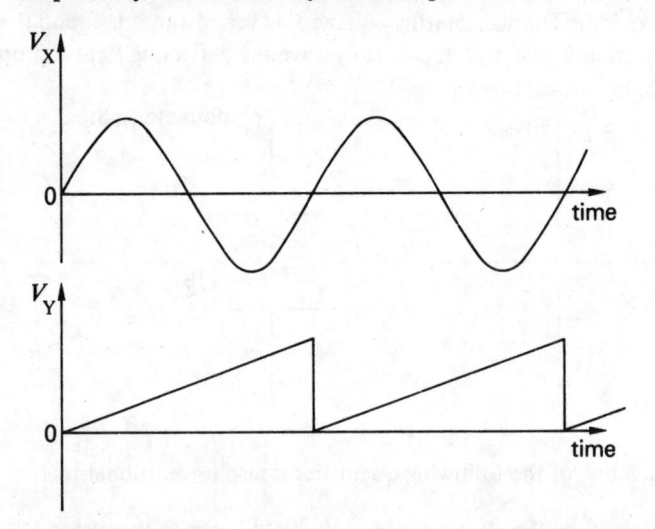

Which one of the following sketches shows the trace appearing on the screen?

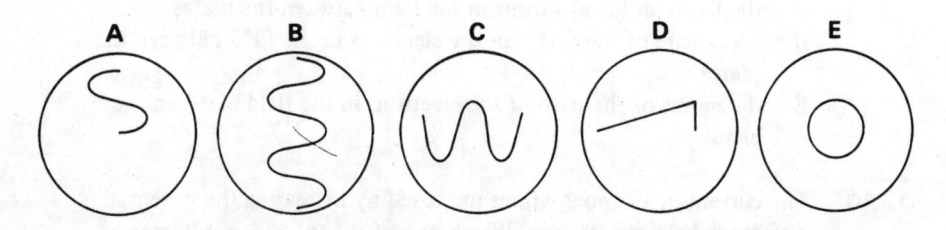

A   B   C   D   E

D 3.8 A particle, of mass m, charge Q and speed v, enters a uniform magnetic field of flux density B and describes a circular path of radius r which is
A   independent of m.
B   inversely proportional to m.
C   inversely proportional to v.
D   directly proportional to Q.
E   inversely proportional to B.

EXERCISE 3

D 3.9 The diagram below represents the cathode, anode and Y deflector plates of a cathode ray tube. There is a constant deflecting electric field between Y_1 and Y_2. The accelerating voltage V is varied and it is found that the deflection θ with which electrons leave the deflecting field is proportional to $1/V$.

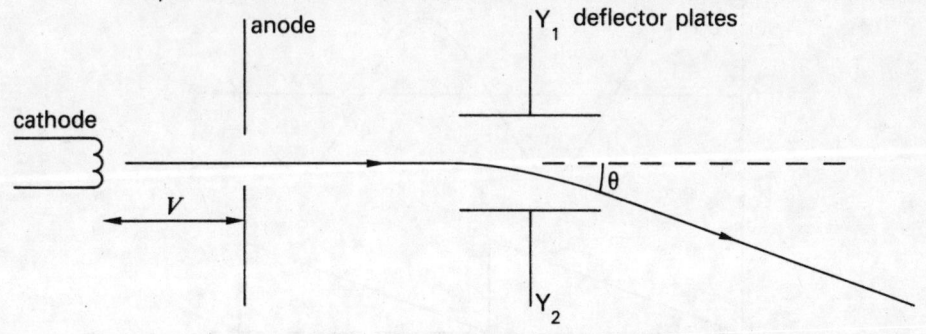

Which one of the following quantities is also proportional to $1/V$?

- **A** the time the electrons are in the field between the plates
- **B** the square of the time the electrons are in the field between the plates
- **C** the force on the electrons in the field between the plates
- **D** the square of the force on the electrons in the field between the plates
- **E** the radius of the path of the electrons in the field between the plates

D 3.10 The current in a copper wire is increased by increasing the potential difference between its ends. Which one of the following statements regarding n, the number of charge carriers per unit volume in the wire, and v, the mean magnitude of the drift velocity of the charge carriers, is correct as the current increases?

- **A** n is unaltered but v is decreased.
- **B** n is unaltered but v is increased.
- **C** n is increased but v is decreased.
- **D** n is increased but v is unaltered.
- **E** Both n and v are increased.

SECTION D

EXERCISE 4

D 4.1 An electron in a vacuum describes a circular path at constant speed. Which of the following drawings, made with the usual conventions, best shows the magnetic flux density pattern in the plane of the path of the electron?

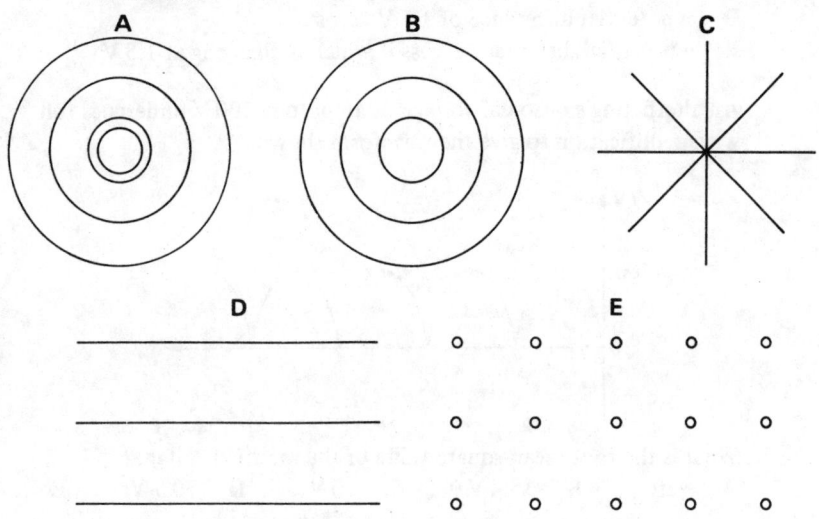

D 4.2 An experimenter investigates the variation of the force F between two long, parallel current-carrying conductors a distance d apart. A straight-line graph should be obtained on plotting
- **A** F against d.
- **B** F against $1/d$.
- **C** F against $1/d^2$.
- **D** $\lg F$ against d.
- **E** F against $\lg d$.

D 4.3 If power P is dissipated when a steady current I flows between two points in a circuit, the potential difference V between the points is defined by P/I. The e.m.f. E induced in a coil by a changing magnetic flux ϕ is given by $E = -d\phi/dt$. It follows that the unit of magnetic flux can be written in SI base units as
- **A** $m\,s^{-2}\,A^{-1}$
- **B** $kg\,m\,s^2\,A^{-1}$
- **C** $m\,s^{-1}\,A$
- **D** $kg\,m^2\,s^{-2}\,A^{-1}$
- **E** $kg\,m^2\,s^{-2}\,A$

EXERCISE 4

D 4.4 Two coils have a mutual inductance of 2.0 H. As a result of mutual induction, an e.m.f. of 3.0 V exists in one coil. It follows that the other coil must have
- A   a current of 1.5 A flowing in it.
- B   a current in it which is changing at 1.5 A s^{-1}.
- C   a charge of 1.5 C passing through it.
- D   a potential difference of 1.5 V across it.
- E   a potential difference across it which is changing at 1.5 V s^{-1}.

D 4.5 An alternating sinusoidal voltage of amplitude 100 V undergoes full wave rectification to give the waveform shown.

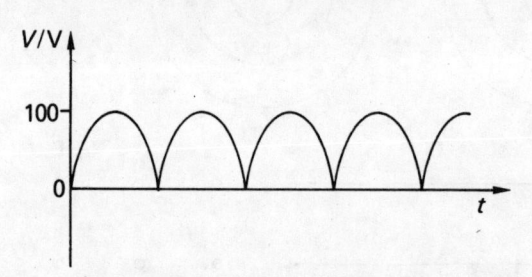

What is the root-mean-square value of the rectified voltage?
   A   zero   B   35.4 V   C   50 V   D   70.7 V   E   100 V

D 4.6 Half-wave rectification of an alternating sinusoidal voltage of amplitude 200 V gives the waveform shown below.

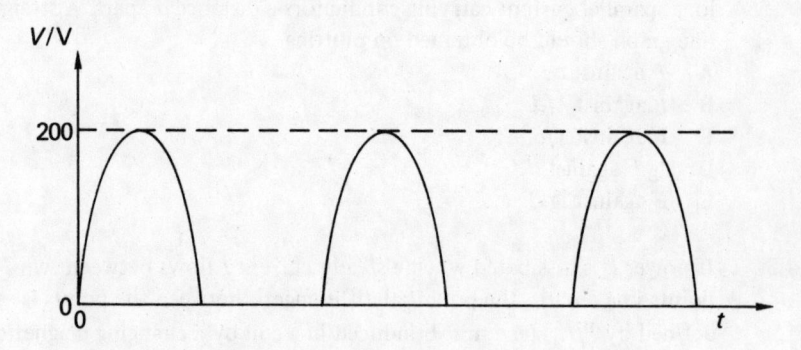

What is the root-mean-square value of the rectified voltage?
   A   zero   B   70.7 V   C   100 V   D   141.4 V
   E   200 V

D 4.7 An alternating potential difference $V_x = V_0 \sin \omega t$ is applied across the X-plates of a cathode-ray oscilloscope, whilst a potential difference $V_y = V_0 \sin(\omega t + \tfrac{1}{2}\pi)$ is applied across the Y-plates. The X-sensitivity is 5/3 times the Y-sensitivity. Which one of the following diagrams represents the trace on the screen?

SECTION D   55

A

B

C

D

E

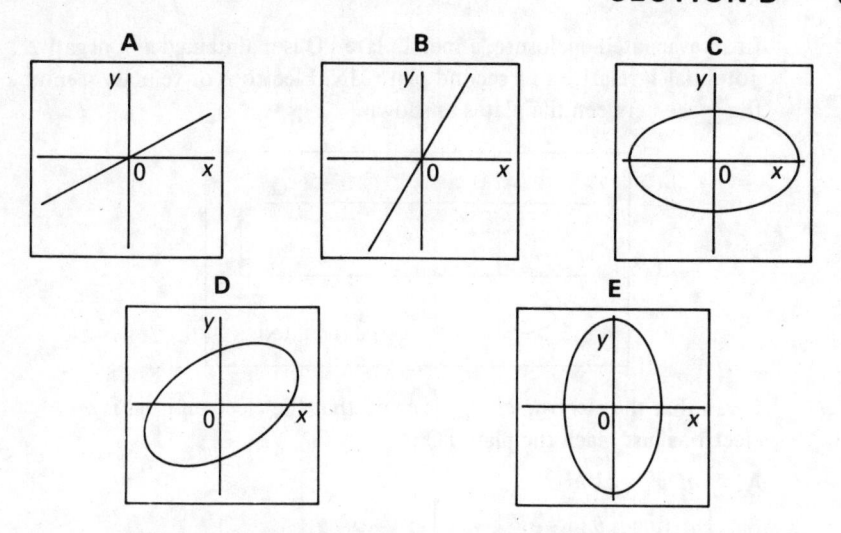

D 4.8   Two particles Y and Z emitted by a radioactive source at P made tracks in a cloud chamber as illustrated in the diagram below. A magnetic field acted downwards into the paper. Careful measurements showed that both tracks were circular, the radius of the Y track being half that of the Z track.

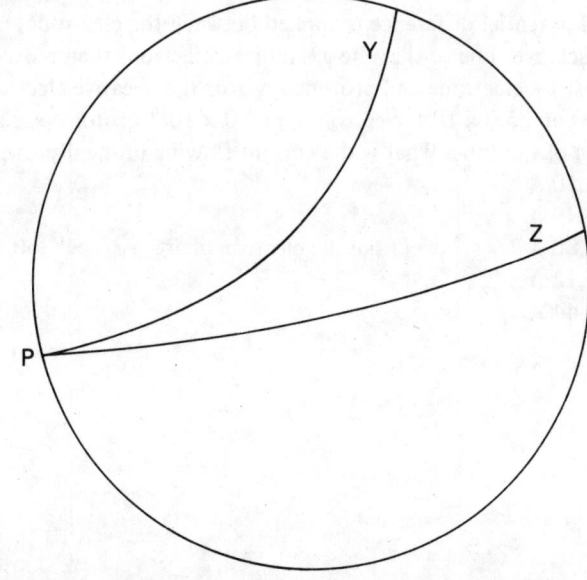

Which one of the following statements is certainly true?
A   The speed of the Z particle was one half that of the Y particle.
B   The mass of the Z particle was one half that of the Y particle.
C   The mass of the Z particle was twice that of the Y particle.
D   The charge of the Z particle was twice that of the Y particle.
E   Both the Y and Z particles carried a positive charge.

EXERCISE 4

D 4.9 In an evacuated enclosure, a metal plate PQ is maintained at a negative potential V relative to a second plate MN. Electrons of velocity v enter the space between the plates as shown.

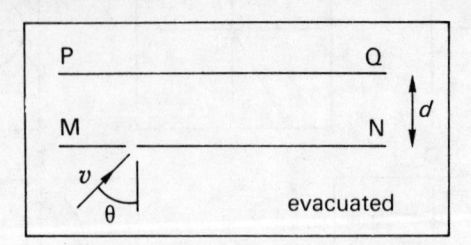

Given that the electron charge is e and that the electron mass is m_e, electrons just reach the plate PQ if

A $\frac{1}{2}m_e v^2 = eV/d$

B $\frac{1}{2}m_e (v \cos \theta)^2 = eV$

C $\frac{1}{2}m_e (v \sin \theta)^2 = eV/d$

D $\frac{1}{2}m_e (v \sin \theta)^2 = eV$

E $\frac{1}{2}m_e (v \cos \theta)^2 = eV/d$

D 4.10 A high potential difference is applied between the electrodes in a hydrogen discharge tube so that the gas ionises; electrons then move towards the positive electrode and protons towards the negative electrode. In each second 5.0×10^{18} electrons and 2.0×10^{18} protons pass a cross-section of the tube. What is the current flowing in the discharge tube?

A   0.10 A
B   0.48 A
C   0.80 A   [Take the electron charge e to be 1.60×10^{-19} C.]
D   1.12 A
E   1.44 A

SECTION E MOLECULAR, THERMAL AND MECHANICAL PROPERTIES OF MATTER

EXERCISE 1

E 1.1 What is the order of magnitude of the diameters of atoms?
    A 10^{-34} m   B 10^{-31} m   C 10^{-19} m   D 10^{-10} m
    E 10^{-7} m

E 1.2 In an experiment to demonstrate Brownian motion in a gas, a brightly illuminated cell containing smoke is viewed under a microscope. The observer sees a large number of bright specks undergoing random, jerky motion. Which one of the following statements about this experiment is correct?
    A Light is being scattered by gas molecules.
    B The motion of gas molecules is being observed directly.
    C The larger the smoke particles, the greater is the speed of the bright specks.
    D The lower the pressure of the gas, the more frequent are the direction changes of the bright specks.
    E The higher the temperature of the gas, the faster is the motion of the bright specks.

E 1.3 The root-mean-square speed of hydrogen molecules is v when at a temperature of 300 K. What is its value at a temperature of 450 K?
    A $v/1.5$ B $v/\sqrt{1.5}$ C $v\sqrt{1.5}$ D $1.5v$
    E $2.25v$

E 1.4 A barometer tube contains a little air in the space above the mercury. The length of the space is 160 mm and the observed height of the mercury is 744 mm when the true atmospheric pressure is 750 mm Hg. When the length of the space is 120 mm and the observed height of the barometer is 770 mm, what is the true atmospheric pressure?
    A 790 mm Hg   B 783 mm Hg   C 778 mm Hg
    D 760 mm Hg   E 748 mm Hg

E 1.5 The temperature at which pure water boils in the open air
    A depends on the rate at which heat is being supplied to it.
    B is always the same at a given height above sea-level.
    C is raised by an increase in the atmospheric pressure.
    D is raised by an increase in the temperature of the air.
    E is lowered by an increase in the relative humidity of the air.

EXERCISE 1

E 1.6 A metal vessel of negligible heat capacity contains 0.50 kg of water which is heated from 15 °C to 85 °C by a 750 W immersion heater in 240 s. What is the average rate of loss of heat to the surroundings from the vessel during this time?
A   70 W   B   123 W   C   138 W   D   212 W
E   375 W
[Take the specific heat capacity of water to be 4.2×10^3 J kg^{-1} K^{-1}.]

E 1.7 The internal energy of a fixed mass of an ideal gas is a function of
A   the pressure, but not of the volume or the temperature.
B   the pressure and the temperature, but not of the volume.
C   the volume, but not of the pressure or the temperature.
D   the volume and the temperature, but not of the pressure.
E   the temperature, but not of the pressure or the volume.

E 1.8 On braking, 500 kJ of heat were produced when a vehicle of total mass 1600 kg was brought to rest on a level road. What was the speed of the vehicle just before the brakes were applied?
A   0.625 m s^{-1} B   0.795 m s^{-1} C   25.0 m s^{-1}
D   62.5 m s^{-1} E   625 m s^{-1}

E 1.9 A lagged copper bar is heated at one end and the other end is in contact with melting ice. If the rate at which heat is supplied to the bar is increased, what happens to the temperature of the hot end, the temperature gradient along the bar and the rate of flow of heat down the bar?

	temperature of hot end	temperature gradient	rate of flow of heat
A	unchanged	unchanged	increases
B	increases	unchanged	increases
C	increases	increases	increases
D	increases	increases	unchanged
E	unchanged	increases	unchanged

E 1.10 Several wires of the same material, but different lengths l and diameters d, are hung vertically from a fixed support and loaded at the lower end until each extends by 1 mm. The magnitude of each of the loads is proportional to
A ld B d/l C l/d^2 D l/d E d^2/l

EXERCISE 2

E 2.1 The number of molecules in 1 dm^3 (litre) of water is approximately
  A 3×10^{19} B 3×10^{21} C 3×10^{23} D 3×10^{25}
  E 3×10^{27}

E 2.2 The pressure of a fixed mass of gas at constant volume is greater at a higher temperature. One reason for this is that
  A the molecules collide with the container walls more frequently.
  B the intermolecular attractions between the gas molecules is less.
  C the mean free path of the molecules is greater.
  D the size of each individual molecule is greater.
  E the energy transferred to the walls during collisions is greater.

E 2.3 A container of fixed volume contains a mass m of ideal gas at pressure p. The root-mean-square speed of the molecules is $c_{r.m.s.}$. A further mass m of the same gas is pumped into the container and the pressure rises to $2p$, the temperature being the same as before. What is now the root-mean-square speed of the molecules?

  A $\dfrac{c_{r.m.s.}}{\sqrt{2}}$ B $c_{r.m.s.}$ C $\sqrt{2}c_{r.m.s.}$ D $2c_{r.m.s.}$

  D $4c_{r.m.s.}$

E 2.4 A hot-air balloon of constant volume contains air at 100 °C. The fraction of this air which escapes if the temperature is increased by 1 K (the pressure remaining constant) is approximately
  A 1/373   B 1/273   C 1/100   D 374/373
  E 101/100

E 2.5 The kelvin, the SI unit of temperature interval, is defined as
  A one-hundredth of the temperature difference between the ice-point and the steam-point.
  B one-hundredth of the temperature difference between the triple-point of water and the steam-point.
  C the fraction 1/273.16 of the thermodynamic temperature of the triple-point of water.
  D the fraction 1/373.15 of the thermodynamic temperature of the steam-point.
  E −273.16 °C.

EXERCISE 2

E 2.6 Two identical copper calorimeters of mass 0.1 kg contain 1 kg of water and 2 kg of alcohol respectively. It is found that they both take 60 minutes to cool from 350 K to 300 K under similar conditions. What is the specific heat capacity of alcohol?

- A 600 J kg^{-1} K^{-1}
- B 2060 J kg^{-1} K^{-1}
- C 2100 J kg^{-1} K^{-1}
- D 2140 J kg^{-1} K^{-1}
- E 3800 J kg^{-1} K^{-1}

[The specific heat capacity of copper is 400 J kg^{-1} K^{-1} and of water is 4200 J kg^{-1} K^{-1}.]

E 2.7 The first law of thermodynamics can be stated in the form $Q = \Delta U + W$. Which of the quantities Q, ΔU, W must necessarily be zero when a *real* gas undergoes a change at constant pressure?

- A Q only
- B ΔU only
- C W only
- D none of Q, ΔU and W
- E all of Q, ΔU and W

E 2.8 A lead bullet of mass 7.0×10^{-3} kg is fired at a target, where its kinetic energy is converted entirely to heat. The initial temperature of the bullet is 27 °C, and the following constants are known:

   specific heat capacity of lead = 140 J kg^{-1} K^{-1};
   specific latent heat of fusion of lead = 2.5×10^4 J kg^{-1};
   melting point of lead = 327 °C.

The heat generated by the impact is absorbed entirely by the bullet, which is just melted. What was the impact velocity of the bullet?

- A 298 m s^{-1}
- B 303 m s^{-1}
- C 366 m s^{-1}
- D 2560 m s^{-1}
- E 67 000 m s^{-1}

E 2.9 A thin-walled metal tank of surface area 5 m^2 is filled with water and contains an immersion heater dissipating 1 kW. The tank is covered with a 4 cm thick layer of insulation whose thermal conductivity is 0.2 W m^{-1} K^{-1}. The outer face of the insulation is at 25 °C. What is the temperature of the tank in the steady state?

- A 29 °C
- B 35 °C
- C 40 °C
- D 65 °C
- E 80 °C

SECTION E

E 2.10 A suspended wire is gradually loaded until it is stretched just beyond its elastic limit, and it is then gradually unloaded. Which of the following graphs (with arrows indicating the sequence) best illustrates the variation of tensile stress σ with tensile strain ε ?

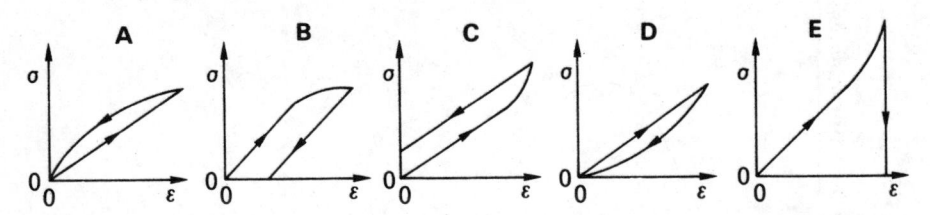

EXERCISE 3

E 3.1 Which of the following is the best estimate of the average spacing between the molecules in a gas at standard temperature and pressure?
  A 0.001 nm   B 0.06 nm   C 3 nm   D 15 nm
  E 300 nm
[The Avogadro constant is 6×10^{23} mol^{-1}.]

E 3.2 The mutual potential energy E_p of two atoms separated by a distance x is shown in the graph below.

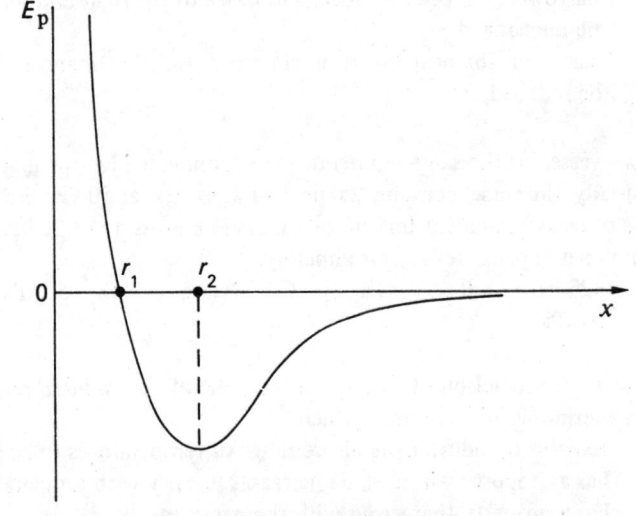

The corresponding force between the atoms is
  A attractive for x less than r_1 and repulsive for x greater than r_1.
  B attractive for x greater than r_1 and repulsive for x less than r_1.
  C attractive for x less than r_2 and repulsive for x greater than r_2.
  D attractive for x greater than r_2 and repulsive for x less than r_2.
  E attractive for all values of x.

E 3.3 The graph below shows the distribution of molecular speeds for gaseous oxygen at room temperature.

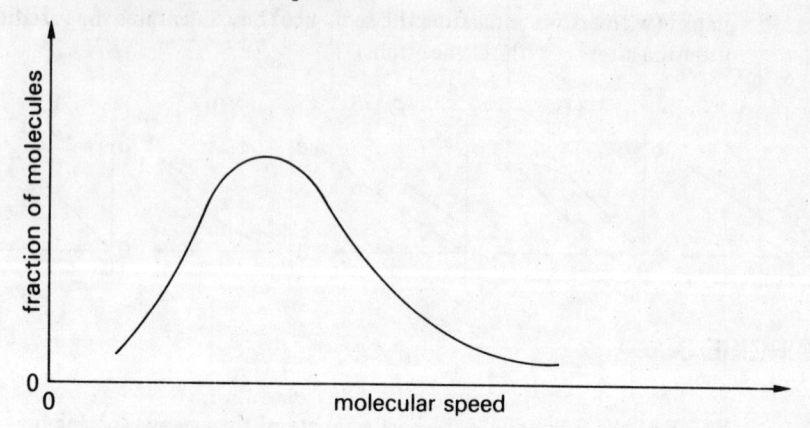

As the temperature of the gas is increased, the curve will become
A   broader, the peak position will move to the right and its height will be unchanged.
B   broader, the peak position will move to the right and its height will be reduced.
C   broader, the peak position will be unchanged and its height will be reduced.
D   narrower, the peak position will move to the right and its height will be unchanged.
E   narrower, the peak position will move to the right and its height will be reduced.

E 3.4 Two vessels at the same temperature are connected by opening a valve. Initially one vessel contains 2.0 dm^3 of argon gas at 50 kPa pressure and the other contains 3.0 dm^3 of xenon gas at a pressure of 20 kPa. The final equilibrium pressure is approximately
A   32 kPa   B   35 kPa   C   38 kPa   D   50 kPa
E   70 kPa

E 3.5 For the construction of a thermometer, one of the *essential* requirements is a thermometric substance which
A   remains liquid over the entire range of temperatures to be measured.
B   has a property whose value increases linearly with temperature.
C   has a property that varies with temperature.
D   obeys Boyle's law.
E   has a constant expansivity.

SECTION E

E 3.6 The temperature of a hot liquid in a container of negligible heat capacity falls at a rate of 2.0 K min^{-1} just before it begins to solidify. The temperature then remains steady for 20 min by which time the liquid has all solidified.

The ratio $\dfrac{\text{specific heat capacity of liquid}}{\text{specific latent heat of fusion}}$ is equal to

A $\frac{1}{40}$ B $\frac{1}{10}$ C 1   D 10   E 40

E 3.7 When a gas expands under reversible isothermal conditions, it does external work. The necessary energy becomes available because
A expansion produces a temperature gradient across the gas.
B heat is supplied to the gas from the surroundings.
C the gas obeys Boyle's law.
D the heat capacity of the gas has been reduced.
E the average kinetic energy of the molecules has been reduced.

E 3.8 The specific heat capacity of copper is determined by rotating a cylindrical copper block of mass m at uniform angular velocity ω against a friction couple C for time t, and observing the rise in temperature ΔT. The value of the specific heat capacity is given by
A $C\omega t/m\Delta T$ B $Ct/m\Delta T$ C $mCt/\omega\Delta T$
D $mC\Delta T/\omega t$ E $C\omega/mt\Delta T$

64 EXERCISE 3

E 3.9 A composite rod of uniform cross-section has copper and aluminium sections of the same length in good thermal contact. The ends of the rod, which is well-lagged, are maintained at 100 °C and at 0 °C as shown in the diagram below. The thermal conductivity of copper is twice that of aluminium.

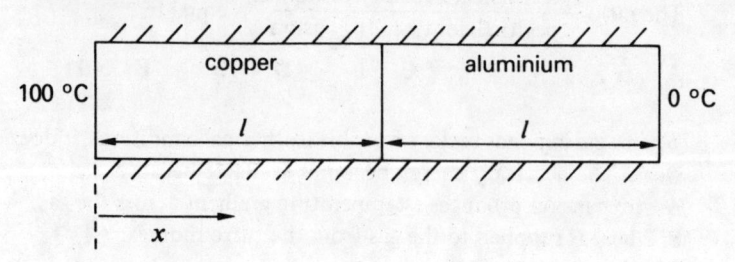

Which of the following graphs best represents the variation of temperature θ with distance x along the rod in the steady state?

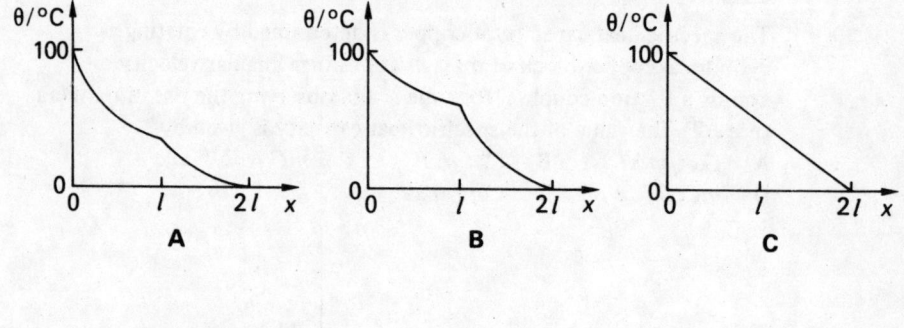

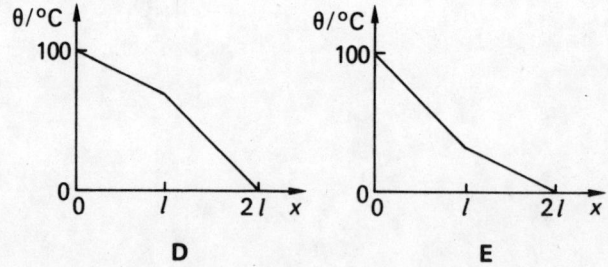

E 3.10 A rod of cross-sectional area 9.0×10^{-6} m^2 is made of a material for which the Young modulus is 2.0×10^{11} Pa (N m^{-2}) and whose breaking strain is 1.0×10^{-3}. Assuming that Hooke's law is obeyed until the rod breaks, what is the tension required to break the rod?

  A   300 N   B   450 N   C   1.8 kN   D   3.6 kN
  E   1.2 MN

EXERCISE 4

E 4.1 The distance between adjacent atoms in a solid is of the same order of magnitude as the wavelength of
A radio waves.
B infrared radiation.
C visible light.
D ultraviolet radiation.
E X-rays.

E 4.2 In the crystal lattice of sodium chloride (NaCl), the principal forces between the lattice particles are
A ionic. B magnetic. C metallic.
D van der Waals. E gravitational.

E 4.3 The speed of nine molecules are distributed as follows:

speed/m s^{-1}	1.0	2.0	3.0	4.0	5.0	6.0
no. of molecules	1	1	4	1	1	1

The root-mean-square speed is approximately
A 1.2 m s^{-1} B 3.0 m s^{-1} C 3.3 m s^{-1}
D 3.6 m s^{-1} E 10.9 m s^{-1}

E 4.4 Two glass bulbs each of internal volume 1 dm^3 are connected by a narrow tube of negligible internal volume. The apparatus is initially filled with air at 27 °C and pressure p. When one of the bulbs is heated to 177 °C, with the second maintained at 27 °C, the resulting pressure in the apparatus becomes
A 1.07p B 1.12p C 1.20p D 1.67p
E 1.80p

66 **EXERCISE 4**

E 4.5　The diagram below shows a constant-volume gas thermometer.

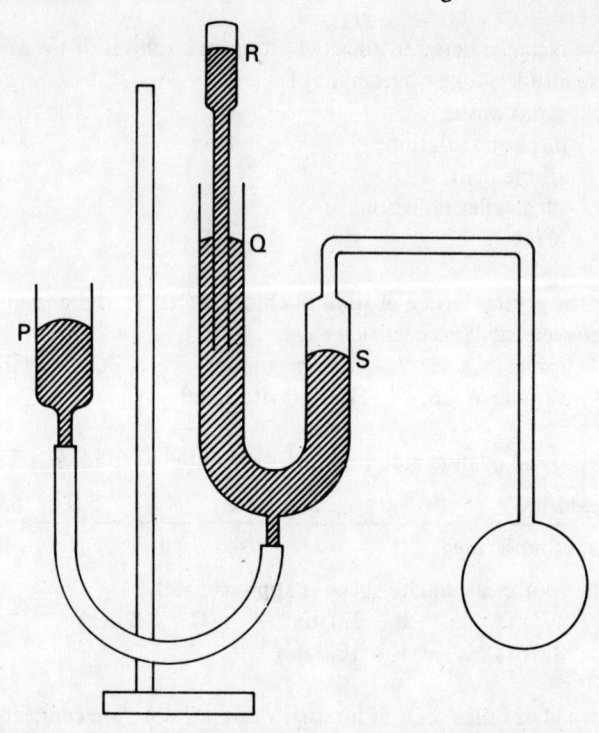

The diagram contains one serious error. What is this error?
A　The mercury at P should be the same level as that at S.
B　R should be open to the atmosphere.
C　The level of the mercury at P should be above that at Q.
D　The tube at P should have the same diameter as the tube at Q.
E　The level of the mercury at P should be the same as that at Q.

SECTION E

E 4.6　A meteorite of mass m is made from solid material of specific heat capacity c, of specific latent heat of fusion l and of very high thermal conductivity. When the meteorite enters the atmosphere from outer space, its temperature is below its melting point by ΔT. Because of atmospheric friction, it absorbs energy at a constant net rate P. The time before the solid becomes completely molten is given by

A　$\dfrac{m(c+l)\Delta T}{P}$

B　$\dfrac{m(c\Delta T + l)}{P}$

C　$m(c+l)\Delta T P$

D　$\dfrac{P}{m\Delta T(c+l)}$

E　$\dfrac{P}{m(c\Delta T + l)}$

E 4.7　1.0 kg of steam at 100 °C occupies 1.65 m³, the specific latent heat of vaporisation of water is 2.26×10^6 J kg^{-1}, and atmospheric pressure is approximately 1.0×10^5 Pa. What proportion of the specific latent heat of vaporisation is utilised for the work associated with expansion when 1 kg of water turns to steam at 100 °C?
A　0.069　　B　0.073　　C　0.105　　D　0.137
E　0.146

E 4.8　In a nuclear reaction, energy equivalent to 1.0×10^{-11} kg of matter is released. This energy is approximately
A　4.5×10^{-6} J　　B　9.0×10^{-6} J　　C　3.0×10^{-3} J
D　4.5×10^5 J　　E　9.0×10^5 J

EXERCISE 4

E 4.9 Two bars of equal length and the same cross-sectional area but of different thermal conductivities, k_1 and k_2, are joined end to end as shown in the diagram. One end of the composite bar is maintained at temperature T_h whereas the opposite end is held at T_c.

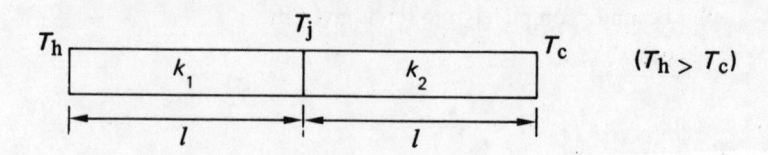

If there are no heat losses from the sides of the bars, the temperature T_j of the junction when steady conditions are attained is given by

A $\dfrac{k_2}{k_1} \cdot \dfrac{(T_h + T_c)}{2}$

B $\dfrac{k_2}{k_1 + k_2} \cdot (T_h + T_c)$

C $\dfrac{k_1 + k_2}{2} \cdot \dfrac{(T_h + T_c)}{2}$

D $\dfrac{1}{(k_1 + k_2)} \cdot (k_1 T_h + k_2 T_c)$

E $\dfrac{(k_1 T_h + k_2 T_c)}{2}$

E 4.10 A wire that obeys Hooke's law is of length x_1 when it is in equilibrium under a tension F_1; its length becomes x_2 when the tension is increased to F_2. The extra energy stored in the wire as a result of this process is

A $\tfrac{1}{4}(F_2 + F_1)(x_2 - x_1)$

B $\tfrac{1}{4}(F_2 + F_1)(x_2 + x_1)$

C $\tfrac{1}{2}(F_2 + F_1)(x_2 - x_1)$

D $\tfrac{1}{2}(F_2 + F_1)(x_2 + x_1)$

E $(F_2 - F_1)(x_2 - x_1)$

SECTION F ATOMIC AND NUCLEAR PROPERTIES

EXERCISE 1

F 1.1 When radiation from a mercury vapour lamp falls on an insulated zinc plate enclosed in a quartz envelope and bearing a negative charge, the plate rapidly loses its charge. This is because
  A the lamp emits positive Hg^{2+} ions which neutralise the charge on the plate.
  B the radiation from the lamp causes electrons to be emitted from the plate.
  C the radiation from the lamp ionises the air and the charge leaks away.
  D the radiation causes positively charged ions to be ejected from the envelope on to the plate.
  E the lamp emits alpha particles, which neutralise the charge.

F 1.2 What is the photoelectric work function of a material?
  A the greatest energy with which an electron can be ejected from a surface by photoelectric emission
  B the least energy with which an electron can be ejected from a surface by photoelectric emission
  C the difference between the energy of an incident photon and the energy of the least energetic electron ejected
  D the difference between the energy of an incident photon and the energy of the most energetic electron ejected
  E the energy of the most energetic photon which will cause photoelectric emission from that surface

F 1.3 What is a photon?
  A the nucleus of a hydrogen atom
  B the particle identical with an electron apart from its positive rather than negative charge
  C the product of the Planck constant and the wavelength of a radiation
  D the quantum of energy associated with an electromagnetic wave
  E the minimum energy needed to remove an electron from a metal surface

F 1.4 The wave nature of electrons is suggested by experiments on
  A line spectra of atoms.
  B the production of X-rays.
  C the photoelectric effect.
  D electron diffraction by a crystalline material.
  E β-decay of nuclei.

70 EXERCISE 1

F 1.5   An electron in a hydrogen atom makes a transition from an energy level with energy E_1 to one with energy E_2 and simultaneously emits a photon. The wavelength of the emitted photon is

  A $h/(E_1 - E_2)$
  B $hc/(E_1 - E_2)$
  C $h/c(E_1 - E_2)$
  D $(E_1 - E_2)/h$
  E $(E_1 - E_2)/hc$

  [where h is the Planck constant and c is the speed of light.]

F 1.6   A beam of γ-rays consists of

  A positively charged particles.
  B negatively charged particles.
  C electromagnetic radiation of wavelength of the order of 10^{-6} m.
  D electromagnetic radiation of wavelength greater than 10^3 m.
  E electromagnetic radiation of wavelength less than 10^{-10} m.

F 1.7   The deviation of α-particles by thin metal foils through angles that range from 0° to 180° can be explained by

  A scattering from free electrons.
  B scattering from bound electrons.
  C diffuse reflection from the metal surface.
  D scattering from small but intense regions of positive charge.
  E diffraction from the crystal lattice.

F 1.8   The decay of $^{238}_{92}$U to $^{239}_{93}$Np by β-emission is not possible because

  A β-decay only occurs in isotopes of low mass.
  B $^{239}_{93}$Np is not a stable isotope.
  C a nucleon number cannot increase in a decay process.
  D a proton number cannot increase in a decay process.
  E a nucleon number and a proton number must both decrease in a decay process.

F 1.9   The rate of decay dN/dt of the number N of nuclei present in a sample of a radioactive element at time t

  A is proportional to t.
  B is proportional to N.
  C is proportional to $1/t$.
  D is proportional to $1/N$.
  E is constant and equal to λ, the decay constant.

SECTION F

F 1.10 If N is the number of radioactive atoms at time t, and $T_{\frac{1}{2}}$ is the half-life, then the rate of change of N with time is

    A $-0.69N/T_{\frac{1}{2}}$ B $+0.69N/T_{\frac{1}{2}}$ C $-N/T_{\frac{1}{2}}$

    D $+N/T_{\frac{1}{2}}$ E $+N$

EXERCISE 2

F 2.1 When the surface of a metal in a vacuum is illuminated by a parallel beam of monochromatic light, which one of the following must be greater than a certain value (depending on the metal) before photo-electrons can be emitted?
- A   the angle of incidence on the surface
- B   the intensity of the beam
- C   the wavelength of the light
- D   the electrical potential of the surface
- E   the frequency of the light

F 2.2 Einstein's equation for the emission of photoelectrons from a metal can be written

$$E = \tfrac{1}{2}m_e v^2 + \phi$$

where E is the energy of a quantum of incident light, and m_e and v stand for the mass and the maximum speed of an emitted photoelectron. What does the symbol ϕ represent in this equation?
- A   the potential energy lost by an electron on leaving the metal
- B   the kinetic energy of an electron ejected from the metal
- C   the energy needed to remove an electron from the K-shell of an atom of the metal
- D   the energy needed to remove one of the most energetic free electrons from the metal
- E   the energy needed to remove one of the least energetic free electrons from the metal

F 2.3 The quantum energy of a photon of yellow light from a sodium lamp is about 3.5×10^{-19} J. Which one of the following would have an energy of about 7×10^{-19} J?
- A   a photon of yellow light from a source that is twice as bright
- B   a photon that consists of two yellow radiations of nearly the same wavelength travelling together
- C   a photon of ultraviolet light
- D   a photon of red light
- E   a photon for which the value of the Planck constant is twice as great

72 EXERCISE 2

F 2.4 Smaller objects may be distinguished by using electron microscopes than by using optical microscopes because
  A electrons are smaller than visible quanta.
  B the electrons travel much faster than light.
  C there is no chromatic aberration with electrons.
  D the electron wavelength is much shorter than that of visible light.
  E the electrons are not diffracted.

F 2.5 The table below shows the six lowest energy levels of the hydrogen atom, labelled $n = 1, 2, 3, \ldots$ etc., with the energy of each level in joules.

  $n = 6$ ——————— -0.6×10^{-19} J
  $n = 5$ ——————— -0.8×10^{-19} J
  $n = 4$ ——————— -1.3×10^{-19} J
  $n = 3$ ——————— -2.4×10^{-19} J
  $n = 2$ ——————— -5.2×10^{-19} J
  $n = 1$ ——————— -22×10^{-19} J

  When an electron moves from $n = 6$ to $n = 1$, the spectrum line emitted has a wavelength of 9.1×10^{-8} m. The wavelength of the spectrum line emitted by the transition from $n = 4$ to $n = 3$ is approximately
  A 4.5×10^{-10} m   B 4.5×10^{-8} m   C 1.8×10^{-6} m
  D 1.8×10^{-4} m   E 1.6×10^{-4} m

F 2.6 Why do alpha particles give bright tracks in a cloud chamber?
  A The alpha particles ionise the air in the chamber and droplets of vapour condense on the ions.
  B The alpha particles are positively charged and act as nuclei on which droplets condense.
  C The gas molecules are excited by the alpha particles and emit flashes of light on returning to a stable state.
  D The chamber contains a plate coated with material which produces visible scintillations when alpha particles strike it.
  E The alpha particles gain electrons from the air molecules to become neutral helium atoms on which water droplets condense.

F 2.7 How many nucleons are there in the nucleus of the atom $^{35}_{17}\text{Cl}$?
  A 17   B 18   C 35   D 52   E 53

F 2.8 Which one of the following combinations of radioactive decay results in the formation of an isotope of the original nucleus?
  A α and four β
  B α and two β
  C α and β
  D two α and β
  E four α and β

F 2.9 A Geiger counter is used to measure the activity of a given radioactive sample. At one instant the meter shows 4000 counts per minute. Five minutes later, it shows 2000 counts per minute. What is the decay constant λ?

A 300ln2   B (ln2)/300   C 300/ln2   D e^2/300
E 300e^2

F 2.10 Taking the half-life of thoron to be 56 s and ln2 to be 0.7 what is the approximate rate of decay when there are 4×10^4 atoms of thoron present?

A 160 Bq   B 320 Bq   C 500 Bq   D 1 000 Bq
E 16 000 Bq

EXERCISE 3

F 3.1 In a photoelectric emission experiment, the maximum kinetic energy of the photoelectrons excited by light of a certain frequency is measured as a function of the intensity of the light. Which of the following graphs best represents the way in which the maximum kinetic energy E_k depends upon the intensity I?

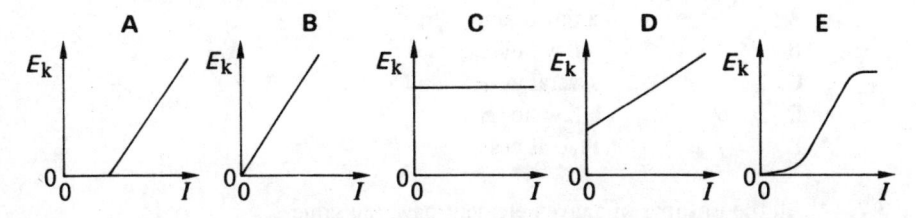

F 3.2 Light quanta of energy 3.5×10^{-19} J fall on the cathode of a photocell. The current through the cell is just reduced to zero by applying a reverse voltage to make the cathode 0.25 V *positive* with respect to the anode. The work function of the metal of the cathode is

A 2.9×10^{-19} J
B 3.1×10^{-19} J
C 3.5×10^{-19} J   [The charge on an electron is -1.6×10^{-19} C.]
D 3.9×10^{-19} J
E 6.4×10^{-19} J

F 3.3 A laser emits monochromatic light of wavelength λ at a constant power P. If h is the Planck constant and c the speed of light in a vacuum, the number of photons emitted per second by the laser is given by

A $\dfrac{Pc}{h\lambda}$ B $\dfrac{\lambda c}{Ph}$ C $\dfrac{hc}{P\lambda}$ D $\dfrac{Ph}{c\lambda}$ E $\dfrac{P\lambda}{hc}$

74 EXERCISE 3

F 3.4 What is the de Broglie wavelength of a rifle bullet of mass 20 g moving at a speed 300 m s^{-1}?

A 7.3×10^{-36} m
B 1.8×10^{-35} m
C 1.1×10^{-34} m   [The Planck constant $h = 6.6 \times 10^{-34}$ J s.]
D 9.9×10^{33} m
E 1.4×10^{35} m

F 3.5 The energies of four levels of the hydrogen atom are as follows:
level P, -13.60 eV;   level Q, -3.40 eV;
level R, -1.50 eV;   level S, -0.85 eV.

Taking the Planck constant as 6.63×10^{-34} J s, the electron charge as -1.60×10^{-19} C and the speed of light as 3.00×10^8 m s^{-1}, a spectral line of 488 nm could result from an electron transition between levels

A Q and P   B R and P   C S and P   D R and Q
E S and Q

F 3.6 In order to trace the line of a water-pipe buried about half a metre below the surface of a field, an engineer proposes to add a radioactive isotope to the water. Which sort of isotope should he choose?

	emitter	*half-life*
A	α	a few hours
B	β	a few hours
C	β	several years
D	γ	a few hours
E	γ	several years

F 3.7 All the isotopes of a given element have the same
A kind of radioactive decay.
B half-life.
C number of nucleons (neutrons + protons) in the nucleus.
D number of neutrons in the nucleus.
E number of protons in the nucleus.

F 3.8 The nucleus of uranium ($^{238}_{92}$U) may undergo successive decays, the total emission consisting of an α-particle, a β-particle and a γ-photon. The resulting nucleus may be represented by

A $^{237}_{92}$U   B $^{234}_{91}$Pa   C $^{233}_{91}$Pa   D $^{232}_{91}$Pa
E $^{234}_{89}$Ac

F 3.9 A pure radioactive nuclide isotope is steadily decaying by a one stage process into a stable nuclide. Which one of the graphs below could represent the decay rate A plotted against the time t?

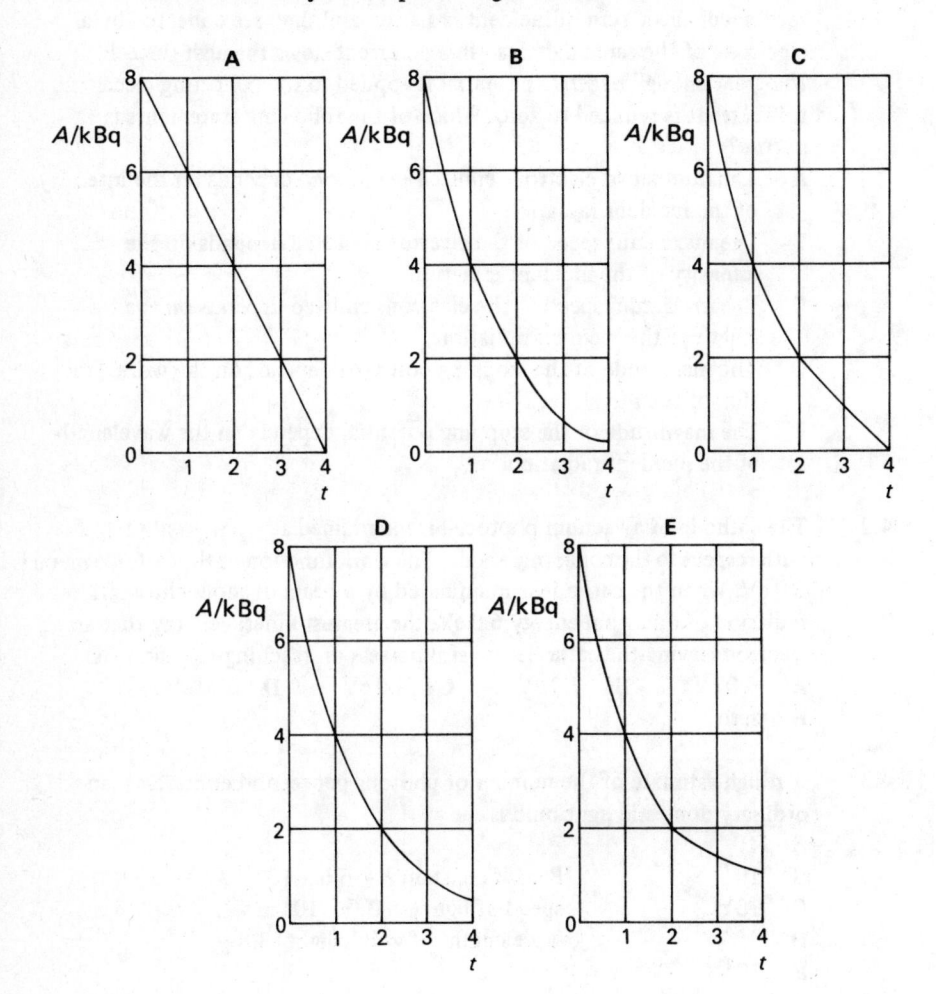

F 3.10 The activity of a radioactive source falls to one-third of its original value in 10 s. What is the half-life of the source?

A $\frac{20}{3}$ s   B $10e^{-\frac{3}{2}}$ C $\frac{10\ln 2}{\ln 3}$ s   D $\frac{5}{\ln 3}$ s

E $10\ln(\frac{2}{3})$ s

EXERCISE 4

F 4.1 In a photoelectric cell, electrons are emitted in vacuo from a metal surface under the action of incident radiation and they are collected by an electrode of the same metal, so that a current flows through the cell.
If a great enough *negative* potential is applied to the collecting electrode, this current is reduced to zero. Which of the following statements is *untrue*?
- **A** The number of electrons emitted per second depends on the intensity of the incident radiation.
- **B** The maximum speed of the electrons emitted depends on the intensity of the incident radiation.
- **C** The maximum speed of the electrons emitted depends on the wavelength of the incident radiation.
- **D** The magnitude of the stopping potential depends on the work function of the metal.
- **E** The magnitude of the stopping potential depends on the wavelength of the incident radiation.

F 4.2 The cathode of a vacuum photocell is maintained at a potential of +1.0 V with respect to the collecting anode. The work function of the cathode metal is 2.0 eV. When the cathode is illuminated by a beam of monochromatic radiation of quantum energy 6.0 eV, the greatest kinetic energy that an electron leaving the cathode can still possess on reaching the anode is
- **A** 9.0 eV **B** 7.0 eV **C** 5.0 eV **D** 3.0 eV
- **E** zero

F 4.3 A rough estimate of the number of photons per second emitted by an ordinary domestic light bulb is
- **A** 10
- **B** 10^7
- **C** 10^{13}
- **D** 10^{20}
- **E** 10^{24}

[Planck constant $h = 6.6 \times 10^{-34}$ J s, speed of light $c = 3.0 \times 10^8$ m s^{-1}, wavelengths of visible light $(3.8$ to $7.2) \times 10^{-7}$ m.]

F 4.4 A beam of light of wavelength λ is totally reflected at normal incidence by a plane mirror. The intensity of the light is such that photons hit the mirror at a rate n. Given that the Planck constant is h, the force exerted on the mirror by this beam is
- **A** $nh\lambda$ **B** nh/λ **C** $2nh\lambda$ **D** $2n\lambda/h$
- **E** $2nh/\lambda$

SECTION F

F 4.5 The energy levels of an electron in a hydrogen atom are given by

$$E_n = \frac{-13.6}{n^2} \text{ eV, where } n = 1, 2, 3, \ldots.$$

The energy required to excite an electron from the ground state to the first excited state is

A   3.4 eV   B   4.5 eV   C   10.2 eV   D   13.6 eV
E   27.2 eV

F 4.6 Why does the nuclide $^{226}_{88}\text{Ra}$ occur in appreciable quantities in a mineral of great geological age, despite the fact that its half-life is only a very small fraction of the age of the mineral?

A   The radioactive decay of $^{226}_{88}\text{Ra}$ is exponential and the activity can never become zero.
B $^{226}_{88}\text{Ra}$ is constantly being formed by the decay of a longer-lived isotope.
C   The mineral is constantly being subjected to neutron bombardment from the centre of the Earth, forming more $^{226}_{88}\text{Ra}$.
D   The concentration of $^{226}_{88}\text{Ra}$ is constantly being regenerated by radio-active fall-out from the atmosphere.
E   The rock surrounding the $^{226}_{88}\text{Ra}$ slows down the escape of the alpha particles and gamma rays.

F 4.7 The binding energy per nucleon may be used as a measure of the stability of a nucleus. This quantity

A   is directly proportional to the neutron/proton ratio of nuclides.
B   is a maximum for nuclides with high nuclear charges.
C   is a maximum for nuclides with medium nuclear masses.
D   is a maximum for nuclides with low nuclear charges.
E   falls to zero for the heaviest radioactive nuclides.

F 4.8 When $^{238}_{92}\text{U}$ is bombarded with slow neutrons it is transformed, absorbing a single neutron and subsequently emitting two β^- particles. The resulting nuclide is represented by

A $^{240}_{93}\text{Np}$ B $^{240}_{91}\text{Pa}$ C $^{239}_{94}\text{Pu}$ D $^{239}_{90}\text{Th}$
E $^{235}_{88}\text{Ra}$

EXERCISE 4

F 4.9 The graph below shows how the number of particles emitted per second A by a radioactive source varies with time t.

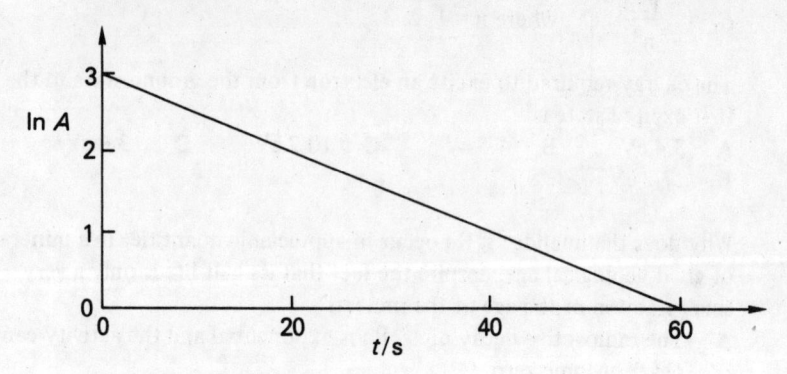

[Graph: ln A vs t/s, straight line from (0, 3) to (60, 0)]

The relationship between A and t is

A $A = 1000\,e^{-(20t)}$

B $A = 20\,e^{(20t)}$

C $A = 3\,e^{-(0.05t)}$ [ln 20 = 3.00]

D $A = 20\,e^{-(0.05t)}$

E $A = 1000\,e^{(0.05t)}$

F 4.10 The $^{14}C : {}^{12}C$ ratio of living material has a constant value during life but the ratio decreases after death because the ^{14}C is not replaced. The half-life of ^{14}C is 5 600 years. The ^{14}C content of a 5.0 g sample of living wood has a radioactive count rate of about 100 per minute. If the count rate of a 10 g sample of ancient wood is 50 per minute, what is the approximate age of the sample?

A 1 400 years B 2 800 years C 5 600 years

D 11 200 years E 22 400 years

SECTION G GRAPHS

G 1 The fundamental frequency f of vibration of a stretched string is given by

$$f = \frac{1}{2l}\sqrt{\frac{T}{m}}$$

where l is the length of the string, T is its tension and m is the mass per unit length of the string. When experimental results for a given string are plotted, which one of the following should give a straight-line graph?

A f plotted against T^2 (l constant)
B f plotted against $2l$ (T constant)
C f^2 plotted against l (T constant)
D f^2 plotted against T (l constant)
E f^2 plotted against $1/T$ (l constant)

G 2 The period of oscillation T of a simple pendulum is measured for known values of l, the length of the string. The gradient of the straight line obtained when T^2 (y-axis) is plotted against l (x-axis) is

A $\dfrac{2\pi}{g}$ B $\dfrac{2\pi}{\sqrt{g}}$ C $\dfrac{4\pi^2}{g}$ D $\dfrac{g}{2\pi}$ E $\dfrac{\sqrt{g}}{4\pi^2}$

G 3 The experimental measurement of the specific heat capacity c of a solid as a function of temperature T is to be fitted to the expression $c = aT + bT^3$. The constants a and b can be found by plotting the straight line graph of

A c against T
B c against T^3
C c against T^2
D $\ln c$ against $\ln T$
E c/T against T^2

80 GRAPHS

G 4 The e.m.f. E of a certain thermocouple is represented by the equation
$$E = a\theta + b\theta^2,$$
where θ is the temperature difference between the two junctions, and a and b are constants. By plotting a graph of $\dfrac{dE}{d\theta}$ (y-axis) against θ (x-axis), a straight line graph will be obtained of slope

A $\dfrac{2b}{a}$ B $\dfrac{b}{a}$ C b D $2b$ E $-2b$

G 5 The magnetic flux density B along the axis of a small bar magnet was measured at various distances x from the centre of the magnet. The results gave the graph shown below.

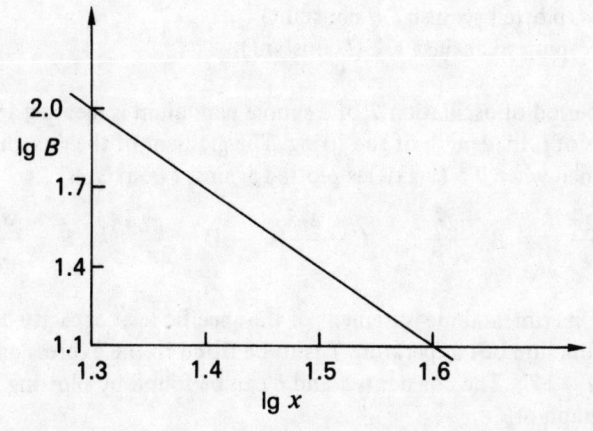

It can be deduced from the graph that B is proportional to
A x^{-3} B x^{-2} C e^{-3x} D e^{2x} E x^2

G 6 The graph below shows how the logarithm of the intrinsic conductivity σ of germanium varies with the reciprocal of temperature T.

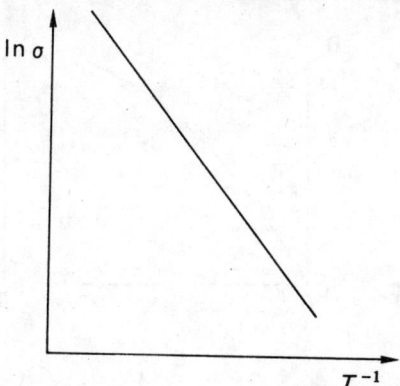

Given that a and b are positive quantities, the graph indicates that the relationship between conductivity and temperature may be put in the form

A $\sigma = T^{-a} + b$
B $\sigma = e^{a/T}$
C $\sigma = be^{-a/T}$
D $\sigma = e^{-a/T} + b$
E $\sigma = T^{-a}$

GRAPHS

Questions G 7 and G 8 refer to the graphs below, each of which shows a quantity y plotted as a function of another quantity x.

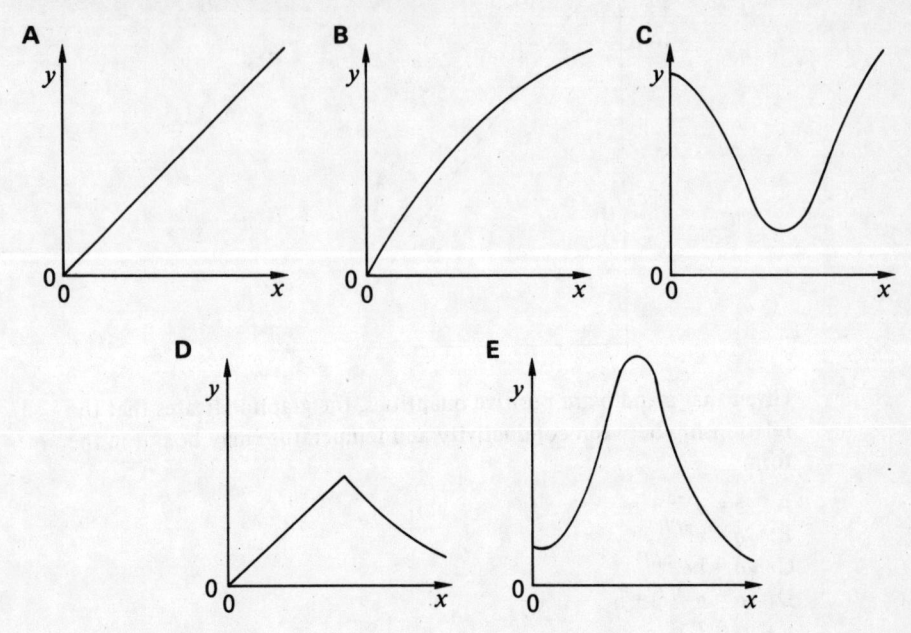

A

B

C

D

E

G 7 Which graph, A, B, C, D, or E, most closely illustrates the correct relationship between the amplitude of oscillation of a pendulum suspended from a vibrating support (y-axis) and the frequency of vibration of the support (x-axis)?

G 8 Which graph, A, B, C, D or E, most closely illustrates the correct relationship between the gravitational force (y-axis) acting on a particle calculated on the assumption that the Earth is a uniform sphere and the distance of the particle from the centre of the Earth (x-axis)?

G 9 The diagram below shows two curves R and S of sinusoidal shape.

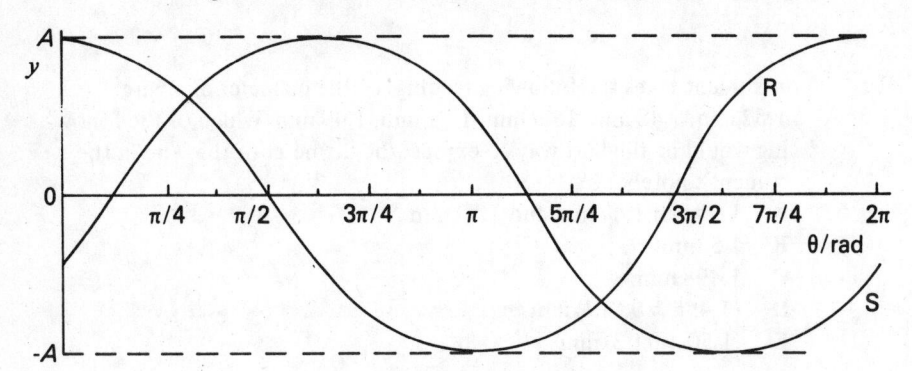

Which one of the following pairs of equations represents these curves?

A $y_R = A \cos\theta$; $y_S = A \sin(\theta - \frac{5}{8}\pi)$
B $y_R = A \cos\theta$; $y_S = A \cos(\theta - \frac{5}{8}\pi)$
C $y_R = A \cos\theta$; $y_S = A \cos(\theta + \frac{5}{8}\pi)$
D $y_R = A \sin\theta$; $y_S = A \sin(\theta - \frac{5}{8}\pi)$
E $y_R = A \sin\theta$; $y_S = A \cos(\theta - \frac{5}{8})$

G 10 An alternating current, $I = I_0 \sin\omega t$, passes through a resistor of resistance R. Which one of the following best represents the variation with time t of the power P dissipated in the resistor?

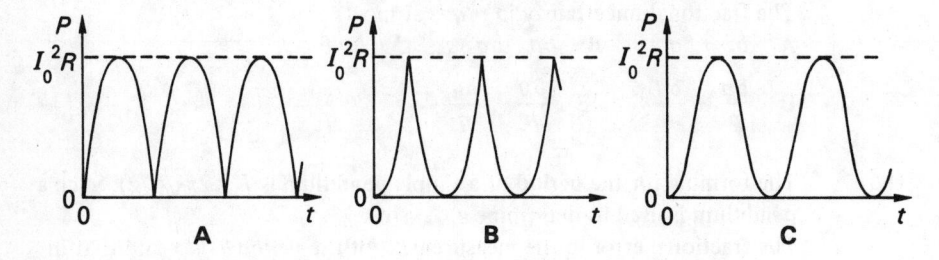

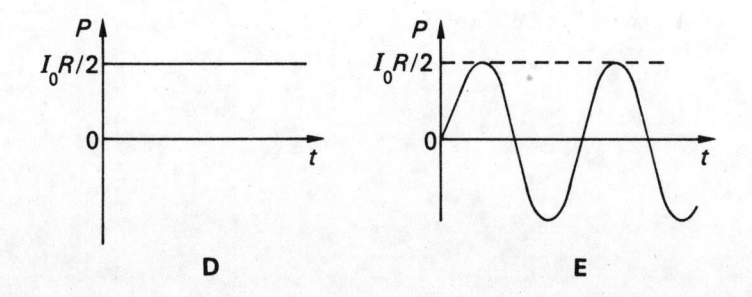

SECTION H ERRORS

H 1 A student takes the following readings of the diameter of a wire: 1.52 mm, 1.48 mm, 1.49 mm, 1.51 mm, 1.49 mm. Which of the following would be the best way to express the diameter of the wire in the student's notebook?
A   between 1.47 mm and 1.53 mm
B   1.5 mm
C   1.498 mm
D   (1.498 ± 0.012) mm
E   (1.50 ± 0.02) mm

H 2 In an experiment, the external diameter d_1 and internal diameter d_2 of a metal tube are found to be (64 ± 2) mm and (47 ± 1) mm respectively. The maximum percentage error in $(d_1 - d_2)$ expected from these readings is approximately
A   0.3%   B   1%   C   5%   D   6%   E   18%

H 3 The dimensions of a rectangular block are measured as (100 ± 1) mm x (80 ± 1) mm x (50 ± 1) mm. The volume of the block calculated from these readings will have a maximum error of about
A $\frac{1}{2}$%   B   1%   C   3%   D   4%   E   6%

H 4 The quantities p and q are measured with estimated errors δp and δq. The fractional uncertainty in p/q is at most
A $\delta p + \delta q$ B $\delta p - \delta q$ C $\delta p . \delta q$
D $\dfrac{\delta p}{p} + \dfrac{\delta q}{q}$ E $\dfrac{\delta p}{p} - \dfrac{\delta q}{q}$

H 5 The formula for the period of a simple pendulum is $T = 2\pi\sqrt{(l/g)}$. Such a pendulum is used to determine g.
The fractional error in the measurement of the period T is x and that in the measurement of the length l is y. The fractional error in the calculated value of g is no greater than
A $x + y$ B $x - y$ C $2x - y$ D $2x + y$
E xy

SECTION H

H 6 In an experiment to measure the speed of sound in free air, the time for sound to pass from a source to a reflector and back to a receiver at the same place as the source is measured. An error of 1% is made in the measurement of the distance of source to reflector and of 2% in the total time. The maximum error in the calculated speed is
A 4% **B** 3% **C** 2% **D** 1% **E** $\frac{1}{2}$%

H 7 A body, dropped from a tower, is timed to take (2.0 ± 0.1) s to fall to the ground. If the acceleration of free fall is agreed to be $10 \, \text{m s}^{-2}$ exactly, the calculated height of the tower should be quoted as
A (20 ± 0.1) m **B** (20 ± 0.2) m **C** (20 ± 0.5) m
D (20 ± 1) m **E** (20 ± 2) m

H 8 The density of a steel ball was determined by measuring its mass and diameter. The mass was measured within 1% and the diameter within 3%. The error in the calculated density of the steel ball is at most
A 2% **B** 4% **C** 8% **D** 10% **E** 28%

H 9 The diameter of the bore of a capillary tube was determined by introducing a small quantity of mercury into the capillary. It was found possible to measure the length of the mercury thread to within 2%, and the mass of the mercury to within 4%. Assuming negligible error in the density of mercury, the error in the calculated diameter of the capillary is at most
A 6% **B** 4% **C** 3% **D** 1% **E** $\frac{1}{2}$%

H 10 The errors in measurements made on a piece of wire are 0.1% for the length, 2% for the diameter and 0.5% for the resistance. The value of the resistivity as calculated from these measurements has a maximum error of
A 0.1% **B** 2.0% **C** 2.4% **D** 4.4% **E** 4.6%

Item Statistics and Correct Answers

For every multiple choice item in this book, the performance of the examination candidates has been analysed to provide a facility value and a discrimination index. These statistics, along with the correct options, are presented in the tables which follow.

The *facility value* of an item is the percentage of candidates who responded correctly (no mark is awarded to any candidate who chooses more than one option). The item's *discrimination index* is the (point-biserial) correlation[1] between success in responding to the item and score on the examination multiple choice test of which it was a part. Thus an item's discrimination index tends to be high when all those who do well on a test respond correctly to the item and when all those who do badly respond incorrectly or not at all. The maximum possible discrimination index is +1 and most of the items in the Advanced Level Physics examinations are intended to have discrimination indices greater than 0.25.

The average facility of the items in this book is approximately 54% and, although much depends on the relationship between the various components of the examination, a score of 54%, or slightly more, on the items in this book is equivalent to a performance at the level of grade D in an A-level examination.

In the table following there will be found, under each item number, the letter for the correct option, the item's facility value and the item's discrimination index.

---

[1] The point-biserial correlation (r) is calculated as follows:

$$r = \frac{M_g - M_t}{\sigma_t} \left(\sqrt{\frac{p}{q}} \right)$$

where M_g = the mean total score on the test of the group choosing the correct option.
M_t = the mean total score on the test of the total sample.
p = the proportion of candidates choosing the correct option.
q = the proportion of candidates not choosing the correct option.
σ_t = the standard deviation of the total test score in the complete sample.

ANSWERS

A1.1	A1.2	A1.3	A1.4	A1.5	A1.6	A1.7	A1.8	A1.9	A1.10
D	B	B	C	E	E	D	A	A	E
65%	48%	53%	72%	76%	36%	82%	36%	31%	46%
0.37	0.24	0.37	0.42	0.41	0.48	0.28	0.21	0.44	0.43
A2.1	A2.2	A2.3	A2.4	A2.5	A2.6	A2.7	A2.8	A2.9	A2.10
B	A	C	D	A	D	C	D	D	A
87%	38%	50%	62%	63%	62%	23%	74%	43%	56%
0.25	0.35	0.14	0.39	0.37	0.41	0.23	0.35	0.29	0.35
A3.1	A3.2	A3.3	A3.4	A3.5	A3.6	A3.7	A3.8	A3.9	A3.10
B	A	D	C	C	D	B	E	A	C
68%	65%	71%	34%	51%	41%	49%	67%	24%	53%
0.39	0.32	0.34	0.29	0.31	0.26	0.38	0.43	0.45	0.41
A4.1	A4.2	A4.3	A4.4	A4.5	A4.6	A4.7	A4.8	A4.9	A4.10
A	E	C	D	A	D	B	B	D	A
43%	78%	55%	52%	18%	52%	59%	50%	47%	51%
0.49	0.38	0.45	0.28	0.24	0.32	0.37	0.41	0.19	0.36
B1.1	B1.2	B1.3	B1.4	B1.5	B1.6	B1.7	B1.8	B1.9	B1.10
C	D	A	E	B	E	D	B	E	D
58%	67%	69%	81%	36%	38%	55%	68%	32%	66%
0.28	0.50	0.31	0.24	0.33	0.40	0.42	0.31	0.37	0.35
B2.1	B2.2	B2.3	B2.4	B2.5	B2.6	B2.7	B2.8	B2.9	B2.10
E	B	B	E	C	B	D	E	D	E
79%	63%	74%	63%	81%	59%	77%	35%	64%	78%
0.42	0.31	0.23	0.42	0.20	0.38	0.39	0.34	0.46	0.37
B3.1	B3.2	B3.3	B3.4	B3.5	B3.6	B3.7	B3.8	B3.9	B3.10
B	D	E	A	B	A	B	C	B	D
78%	46%	32%	45%	74%	77%	57%	88%	51%	75%
0.42	0.50	0.37	0.37	0.40	0.28	0.30	0.26	0.39	0.35
B4.1	B4.2	B4.3	B4.4	B4.5	B4.6	B4.7	B4.8	B4.9	B4.10
C	D	D	A	C	D	A	B	D	C
51%	87%	48%	37%	59%	32%	38%	29%	52%	30%
0.51	0.34	0.42	0.41	0.39	0.21	0.37	0.31	0.31	0.16
C1.1	C1.2	C1.3	C1.4	C1.5	C1.6	C1.7	C1.8	C1.9	C1.10
E	A	A	E	D	B	B	C	C	C
89%	70%	48%	45%	51%	49%	63%	57%	69%	37%
0.28	0.38	0.31	0.39	0.38	0.42	0.28	0.46	0.31	0.51

ANSWERS

C2.1	C2.2	C2.3	C2.4	C2.5	C2.6	C2.7	C2.8	C2.9	C2.10
A	C	B	D	C	D	D	B	D	C
42%	56%	43%	67%	83%	69%	35%	54%	82%	15%
0.43	0.50	0.48	0.44	0.36	0.31	0.39	0.30	0.36	0.16

C3.1	C3.2	C3.3	C3.4	C3.5	C3.6	C3.7	C3.8	C3.9	C3.10
C	C	B	C	C	D	C	E	A	B
51%	59%	40%	56%	54%	51%	33%	51%	61%	57%
0.48	0.46	0.30	0.42	0.32	0.40	0.37	0.33	0.47	0.42

C4.1	C4.2	C4.3	C4.4	C4.5	C4.6	C4.7	C4.8	C4.9	C4.10
A	C	C	B	C	C	C	D	D	D
33%	67%	20%	64%	71%	54%	65%	41%	55%	43%
0.28	0.34	0.27	0.51	0.35	0.41	0.39	0.41	0.46	0.43

D1.1	D1.2	D1.3	D1.4	D1.5	D1.6	D1.7	D1.8	D1.9	D1.10
D	A	E	D	C	C	D	C	D	D
23%	65%	77%	85%	48%	39%	63%	59%	21%	35%
0.30	0.31	0.38	0.34	0.38	0.34	0.30	0.44	0.31	0.50

D2.1	D2.2	D2.3	D2.4	D2.5	D2.6	D2.7	D2.8	D2.9	D2.10
B	D	B	D	C	C	D	E	B	C
68%	72%	36%	53%	46%	46%	42%	35%	58%	58%
0.27	0.29	0.32	0.38	0.42	0.47	0.29	0.48	0.51	0.42

D3.1	D3.2	D3.3	D3.4	D3.5	D3.6	D3.7	D3.8	D3.9	D3.10
E	E	E	B	C	A	A	E	B	B
41%	41%	54%	32%	46%	83%	17%	73%	26%	56%
0.18	0.41	0.38	0.35	0.53	0.36	0.20	0.44	0.33	0.38

D4.1	D4.2	D4.3	D4.4	D4.5	D4.6	D4.7	D4.8	D4.9	D4.10
E	B	D	B	D	C	C	E	B	D
40%	39%	54%	64%	77%	9%	50%	29%	41%	26%
0.39	0.31	0.47	0.46	0.37	−0.05	0.35	0.38	0.36	0.26

E1.1	E1.2	E1.3	E1.4	E1.5	E1.6	E1.7	E1.8	E1.9	E1.10
D	E	C	C	C	C	E	C	C	E
59%	69%	48%	59%	81%	72%	48%	59%	55%	44%
0.41	0.37	0.42	0.38	0.29	0.43	0.45	0.39	0.28	0.43

E2.1	E2.2	E2.3	E2.4	E2.5	E2.6	E2.7	E2.8	E2.9	E2.10
D	A	B	A	C	C	D	C	D	B
35%	74%	40%	44%	73%	83%	42%	52%	58%	78%
0.38	0.24	0.37	0.33	0.30	0.29	0.37	0.48	0.35	0.21

ANSWERS

E3.1	E3.2	E3.3	E3.4	E3.5	E3.6	E3.7	E3.8	E3.9	E3.10
C	D	B	A	C	A	B	A	D	C
28%	50%	44%	61%	35%	48%	44%	66%	47%	91%
0.21	0.43	0.34	0.36	0.27	0.39	0.32	0.44	0.41	0.28

E4.1	E4.2	E4.3	E4.4	E4.5	E4.6	E4.7	E4.8	E4.9	E4.10
E	A	D	C	E	B	B	E	D	C
80%	88%	50%	40%	38%	52%	35%	64%	65%	53%
0.26	0.19	0.40	0.37	0.23	0.40	0.33	0.48	0.54	0.28

F1.1	F1.2	F1.3	F1.4	F1.5	F1.6	F1.7	F1.8	F1.9	F1.10
B	D	D	D	B	E	D	C	B	A
66%	26%	80%	58%	68%	76%	70%	76%	53%	53%
0.36	0.40	0.36	0.40	0.44	0.35	0.42	0.33	0.48	0.47

F2.1	F2.2	F2.3	F2.4	F2.5	F2.6	F2.7	F2.8	F2.9	F2.10
E	D	C	D	C	A	C	B	B	C
66%	39%	40%	60%	43%	70%	72%	70%	56%	67%
0.46	0.27	0.33	0.22	0.45	0.17	0.38	0.42	0.39	0.37

F3.1	F3.2	F3.3	F3.4	F3.5	F3.6	F3.7	F3.8	F3.9	F3.10
C	B	E	C	E	D	E	B	D	C
45%	36%	40%	69%	33%	64%	68%	77%	59%	48%
0.51	0.43	0.44	0.39	0.44	0.31	0.35	0.40	0.41	0.39

F4.1	F4.2	F4.3	F4.4	F4.5	F4.6	F4.7	F4.8	F4.9	F4.10
B	D	D	E	C	B	C	C	D	D
54%	43%	37%	45%	40%	45%	15%	58%	52%	33%
0.46	0.23	0.31	0.38	0.45	0.35	0.30	0.40	0.34	0.24

G1	G2	G3	G4	G5	G6	G7	G8	G9	G10
D	C	E	D	A	C	E	D	B	C
66%	86%	60%	66%	35%	38%	35%	40%	39%	31%
0.44	0.31	0.47	0.41	0.43	0.27	0.29	0.39	0.26	0.21

H1	H2	H3	H4	H5	H6	H7	H8	H9	H10
E	E	D	D	D	B	E	D	C	E
59%	42%	65%	64%	65%	49%	23%	61%	39%	52%
0.24	0.36	0.30	0.29	0.41	0.15	0.37	0.44	0.46	0.38